Management and Industrial Engineering

Series Editor

J. Paulo Davim, Department of Mechanical Engineering, University of Aveiro, Aveiro, Portugal

This series fosters information exchange and discussion on management and industrial engineering and related aspects, namely global management, organizational development and change, strategic management, lean production, performance management, production management, quality engineering, maintenance management, productivity improvement, materials management, human resource management, workforce behavior, innovation and change, technological and organizational flexibility, self-directed work teams, knowledge management, organizational learning, learning organizations, entrepreneurship, sustainable management, etc. The series provides discussion and the exchange of information on principles, strategies, models, techniques, methodologies and applications of management and industrial engineering in the field of the different types of organizational activities. It aims to communicate the latest developments and thinking in what concerns the latest research activity relating to new organizational challenges and changes world-wide. Contributions to this book series are welcome on all subjects related with management and industrial engineering. To submit a proposal or request further information, please contact Professor J. Paulo Davim, Book Series Editor, pdavim@ua.pt

Carolina Machado
Editor

Building the Future with Human Resource Management

Springer

Editor
Carolina Machado
Department of Management
School of Economics and Management
University of Minho
Braga, Portugal

ISSN 2365-0532 ISSN 2365-0540 (electronic)
Management and Industrial Engineering
ISBN 978-3-031-52810-1 ISBN 978-3-031-52811-8 (eBook)
https://doi.org/10.1007/978-3-031-52811-8

This Springer imprint is published by the registered company Springer Nature Switzerland AG
The registered company address is: Gewerbestrasse 11, 6330 Cham, Switzerland

Paper in this product is recyclable.

Preface

The world is in continuous transformation. Innovation, creativity, emotional intelligence, technology, digitalization, artificial intelligence, entrepreneurship and sustainability, among others, are concepts that increasingly circulate among us, demonstrating the challenges and changes that different economies and societies are facing. Common to all these concepts are people, understood as the key and fundamental element capable of creating and leading the world toward a more conscious and sustainable future. In this "new" world, there is an integration between intelligent automation, devices and systems in the workplace in order to increase cooperation and collaboration between man and machine. High-skilled workers thus achieve greater abilities to guide machines and other intelligent technologies, working better and faster with collaborative technologies. Work thus becomes increasingly talent. The stimulation and application of human intelligence contribute to the promotion of organizations that are increasingly inspiring, collaborative and committed to creating a work context compatible with higher levels of human creativity and intellectual power, with a view to achieving greater organizational efficiency and effectiveness.

Building the future with HRM is based on more innovative and creative organizations, with a strong sense of ethics and social responsibility. Only socially responsible people management allows the organization to contribute to the achievement of higher levels of sustainability. Committed to sustainability, organizations seek to improve the quality of life of all those who compose it and who are related to it, in the present as well as future generations. Effectively, with a particular focus on people, the business, and its internal and external environments, organizations are more capable of facing and overcoming more efficiently and effectively the challenges they face not only in the present, but also and increasingly more in the future without compromising the ability of future generations to meet their own challenges. The alignment of issues related to sustainability when defining the organizational strategy and the adequate integration of the different stakeholders of the organization, with a view to ethical and responsible performance, which is concerned with the current business and future generations, are at the base of sustainable organizations. For these, the reduction of the ecological footprint is fundamental. At this level, digitization becomes crucial not only to reduce or even eliminate physical processes that are not critical to the organization, but also to lead to the implementation of new ways of working (such as remote work), and modern management systems, more compatible with the identification and use of opportunities for improvement in real time.

Conscious of these challenges, with this book, entitled *Building the Future with Human Resource Management*, we look to present the different contributions that have been made for the construction of organizations of the future, with a particular emphasis on the role that the people who are part of them assume in this evolutionary context and subject to profound changes and constant challenges. It looks to provide managers, leaders as well as all kinds of professionals with the most recent and critical knowledge and abilities/competencies, to lead with the processes, tools and mechanisms able to build the future of our organizations through and with all of their stakeholders, with a particular focus on the people that constitute its differentiating factor and enhancer of achieving a more effective organizational success. *Building the Future with Human Resource Management* looks to be a huge contribution to organizations to achieve concrete and competitive results as well as to be prepared for the new and continuous challenges. To this end, different experts from different fields of research, in management and engineering, throughout the various chapters that constitute this book, report on the strategies, models, practices and tools able to provide the executives, managers and engineers with the knowledge and good practices needed to organize, manage, implement and control the appropriate programs needed to build their organizations' future with their HRM.

Contributing to stimulate the growth and development of each individual in a dynamic, competitive, global, responsible and sustainable economy, this book, organized into nine chapters, starts by highlighting "Technological Trends in Human Resource Management—Innovation Analysis". Followed by "Artificial Intelligence (AI) in Human Resource Management (HRM)"; the next chapter discusses "Should Human Resources Management "Go Green"?: The Impact of Green Human Resources Management on Employees' Green Behavior, Affective Commitment and Company Green Performance". It follows an approach to "The Design of the Cyberphysical Navigator Model as a Sustainable Framework Enhancing Organisational Resilience"; analyzing the following chapter the "Responsible Management: Structural Components of Business Ethics". "Responsible Human Resource Management: A Strategic Approach" is also a target of analysis, presenting the next chapter "Ideological PC in HRM: Learning with Military". Focusing the next chapter on "Design of Cognition, Cognition of Design—A Cybernetics and Phenomenological Approach to Enhancing Organizational Cognition", the book ends with an approach to "Promoting Gender Equality in Companies: A Path Forward in Building a More Responsible Future—Diagnosis of the "*Happy Company Group*"".

Building the Future with Human Resource Management can be used by a variety of potential stakeholders, including academics/researchers, managers, engineers, practitioners and other professionals in the different areas of business, management and engineering. Students from different undergraduate/graduate levels (undergraduate, masters and doctoral/Ph.Ds), from the areas of management and industrial engineering, are also potential stakeholders, who need to know and understand the challenges and changes that organizations are facing nowadays, namely (and just for example) at the level of digitalization, sustainability, innovation, creativity and entrepreneurship, obtaining, in this way, the knowledge necessary to build more responsible and sustainable organizations, as well as a more sustainable society, for present and future generations.

The editor acknowledges her gratitude to Springer for this opportunity and for its professional support. Finally, we would like to thank all the chapter authors for their interest and availability to work on this project.

Braga, Portugal Carolina Machado

Contents

Technological Trends in Human Resource Management—Innovation Analysis 1
Dalia Suša Vugec, Jasmina Pivar, and Ana-Marija Stjepić

Artificial Intelligence (AI) in Human Resource Management (HRM) 37
Hafinas Halid, Kamalesh Ravesangar, Syaza Lyana Mahadzir, and Siti Norjannah Abd Halim

Should Human Resources Management "Go Green"?: The Impact of Green Human Resources Management on Employees' Green Behavior, Affective Commitment and Company Green Performance 71
Guida Dias, Neuza Ribeiro, Daniel Roque Gomes, Alexandra Leandro, and Maria João Santos

The Design of the Cyberphysical Navigator Model as a Sustainable Framework Enhancing Organisational Resilience 95
Qeis Kamran and Marcus Becker

Responsible Management: Structural Components of Business Ethics 127
Mine Afacan Findikli

Responsible Human Resource Management: A Strategic Approach 147
Santiago Gutiérrez-Broncano, Mercedes Rubio-Andrés, Pedro Jiménez-Estévez, John Opute, and Camilo Giraldo-Giraldo

Ideological PC in HRM: Learning with Military 169
Laurentino Guimarães, Ana Paula Ferreira, and Gina Gaio Santos

Design of Cognition, Cognition of Design—A Cybernetics and Phenomenological Approach to Enhancing Organizational Cognition ... 191
Qeis Kamran

Promoting Gender Equality in Companies: A Path Forward in Building a More Responsible Future—Diagnosis of the "*Happy Company Group*" ... 229
Luís Miguel Martins Sá and Carolina Feliciana Machado

Index ... 245

Editors and Contributors

About the Editors

Carolina Machado received her Ph.D. degree in Management Sciences (Organizational and Policies Management area/Human Resources Management) from the University of Minho in 1999, Master's degree in Management (Strategic Human Resource Management) from Technical University of Lisbon in 1994 and Degree in Business Administration from University of Minho in 1989. Teaching Human Resources Management subjects since 1989 at University of Minho, she is since 2004 Associate Professor (with Habilitation since 2022), with experience and research interest areas in the field of Human Resource Management, International Human Resource Management, Human Resource Management in SMEs, Training and Development, Emotional Intelligence, Management Change, Knowledge Management and Management/HRM in the Digital Age/Business Analytics, Sustainability and Higher Education for Sustainability. She is Head of the Human Resources Management Work Group at the School of Economics and Management at University of Minho, Coordinator of Advanced Training Courses at the Interdisciplinary Centre of Social Sciences, Member of the Interdisciplinary Centre of Social Sciences (CICS.NOVA.UMinho), University of Minho, as well as Chief Editor of the International Journal of Applied Management Sciences and Engineering (IJAMSE), She is Guest Editor of journals, book Editor and book Series Editor, as well as reviewer in different international prestigious journals. In addition, she has also published both as editor/co-editor and as author/co-author several books, book chapters and articles in journals and conferences. e-mail: carolina@eeg.uminho.pt

Contributors

Marcus Becker International School of Management (ISM), Dortmund, Germany

Guida Dias Center of Applied Research on Management and Economics—CARME, School of Technology and Management, Polytechnic Institute of Leiria, Leiria, Portugal

Ana Paula Ferreira School of Economics and Management and CICS NOVA UMinho Research Centre, University of Minho, Braga, Portugal

Mine Afacan Findikli FEASS, Business Department, Istinye University, İstanbul, Turkey

Camilo Giraldo-Giraldo Faculty of Law and Social Sciences, University of Castilla-La Mancha, Toledo, Spain

Daniel Roque Gomes School of Education, Polytechnic Institute of Coimbra, Coimbra, Portugal;
Research Centre for Natural Resources Environment and Society (CERNAS), Coimbra, Portugal

Laurentino Guimarães Higher School of Technology of Fafe and CIDI-IEES, European Institute of Higher Studies, Medelo, Portugal

Santiago Gutiérrez-Broncano Faculty of Social Sciences and Information Technologies, Business Administration Department, University of Castilla-La Mancha, Talavera de la Reina, España

Hafinas Halid School of Business Management, Universiti Utara Malaysia, Kedah, Malaysia

Siti Norjannah Abd Halim Albukhary International University, Alor Setar, Kedah, Malaysia

Pedro Jiménez-Estévez Faculty of Law and Social Sciences, Business Administration Department, University of Castilla-La Mancha, Toledo, Spain

Qeis Kamran International School of Management (ISM), Dortmund, Germany;
Department of Engineering Technology, University of Twente, Enschede, The Netherlands;
Weiden Business School, Ostbayerische Technische Hochschule Amberg-Weiden, Weiden in der Oberpfalz, Germany

Alexandra Leandro School of Education, Polytechnic Institute of Coimbra, NICSH—Research Group in Social and Human Sciences University of Minho—CECS—Communication and Society Research Center, Coimbra, Portugal

Carolina Feliciana Machado School of Economics and Management, University of Minho, Braga, Portugal;
Interdisciplinary Centre of Social Sciences (CICS.NOVA.UMinho), University of Minho, Braga, Portugal

Syaza Lyana Mahadzir Albukhary International University, Alor Setar, Kedah, Malaysia

John Opute Innovation, Leadership, Strategy and Management Department, London South Bank University, London, UK

Jasmina Pivar Faculty of Economics and Business, University of Zagreb, Zagreb, Croatia

Kamalesh Ravesangar Tunku Abdul Rahman University of Management and Technology (TARUMT), Kuala Lumpur, Penang, Malaysia

Neuza Ribeiro Center of Applied Research on Management and Economics—CARME, School of Technology and Management, Polytechnic Institute of Leiria, Leiria, Portugal

Mercedes Rubio-Andrés Faculty of Comerce, Complutense University of Madrid, Madrid, Spain

Luís Miguel Martins Sá School of Economics and Management, University of Minho, Braga, Portugal

Gina Gaio Santos School of Economics and Management and CICS NOVA UMinho Research Centre, University of Minho, Braga, Portugal

Maria João Santos Lisbon School of Economics and Management—ISEG/Ulisboa, SOCIUS—Centro de Investigação em Sociologia e das Organizações, Lisbon, Portugal

Ana-Marija Stjepić Faculty of Economics and Business, University of Zagreb, Zagreb, Croatia

Dalia Suša Vugec Faculty of Economics and Business, University of Zagreb, Zagreb, Croatia

Technological Trends in Human Resource Management—Innovation Analysis

Dalia Suša Vugec, Jasmina Pivar, and Ana-Marija Stjepić

Abstract Human Resource Management (HRM) plays one of the crucial roles in achieving productivity and success in modern organizations. Competent human resources are very important for driving research, innovation, and patent creation. Nowadays, there is a growing interest in innovative HRM practices and their connection to technological advancements. Hence, this study presents a comprehensive analysis of innovation in the HRM field using patent data. The methodology involved extensive data collection, patent landscape analysis, and an exploration of the technical content of HRM patents. The analysis is based on a dataset of 2109 patent families related to HRM using PatSeer databases. The findings shed light on the geographical distribution of HRM innovation, temporal trends, key players, technical diversity, core innovation themes, and technology intersections, and thus provided a foundation for understanding HRM innovation from a patent-driven perspective. While providing useful insights, it also emphasizes the need for more study to investigate the practical consequences, partnerships, and changing trends in the dynamic subject of HRM.

Keywords HRM · Patent analysis · Patent landscape · HRM innovation · Patents

1 Introduction

Since workers make up a significant portion of any organization, human resource management (HRM) is the discipline of research and practice that focuses on people in the workplace (Stewart & Brown, 2019). As a result, it stands to reason that businesses with more productive staff members are more likely to succeed (Stewart & Brown, 2019). Therefore, the management of human resources has consistently been a hot topic in both academic and practical areas (e.g. Anwar & Abdullah, 2021;

D. Suša Vugec (✉) · J. Pivar · A.-M. Stjepić
Faculty of Economics and Business, University of Zagreb, Zagreb, Croatia
e-mail: dsusa@efzg.hr

C. Machado (ed.), *Building the Future with Human Resource Management*,
Management and Industrial Engineering, https://doi.org/10.1007/978-3-031-52811-8_1

Mathis et al., 2015; Stone et al., 2020; Wood, 1999). Moreover, many authors nowadays are concentrating on creative HRM and connecting it with various technological advancements (e.g. Azizi et al., 2021; Vrontis et al., 2022; Zeng & Qi, 2021). Today's nations have relied heavily on their competent human resources to advance their growth and welfare. These resources include thinking, questioning, conducting innovation, generating technology, and thinking. Without efficient human resources, research and innovation would not exist. Only through information and its dissemination is the advancement of research and innovation possible. Patent creation, publishing, and creative works are ways that knowledge manifests itself (Bircan & Gencler, 2015).

If one refers to innovation, in general, it usually means something that's new, unfamiliar, and different from what's considered normal. The advancement of humanity relies on the introduction and implementation of ideas that were previously conceived and brought to life. Without ideas and real inventions, it would be impossible to achieve social growth. Innovations impact every individual, company, nation, and the entire planet. They bring about changes in the world that have captured the attention of both scientists and professionals. The European Commission (1995), Jolly (2008), and Radu (2015) define innovation as an innovative solution to economic or social issues that are focused on the needs of individuals or society as a whole and whose use contributes to the change of the global economy.

A patent is frequently used to protect an answer to a problem by which the patent owner of the invention has exclusive rights for making, modifying, using, and selling it (Bosilj Vukšić, 2020; Ma, 2015). The validity of a patent typically lasts up to 20 years, although there are exceptions for fields where the protection can be extended (Bosilj Vukšić, 2020; wipo.int, 2023). However, the patent holder is normally required to pay specific fees regularly during that time for the invention to stay valid (wipo.int, 2023). Various technologies can be patented, including technologies within the pharmaceutical industry, medicine (medicine devices), biotechnology, chemistry, mechanics, energy-related technologies, automotive-related technologies, optics, electronics, computer-related technologies such as hardware and software, etc. (Allison & Lemley, 2000). It should be noted that software patents in the European Union are only granted in circumstances (Bosilj Vukšić, 2020).

Furthermore, it is also important to note that there is a difference between the terms "invention" and "innovation". Namely, a certain invention must be produced, adequately promoted through marketing activities and channels, and distributed to end customers to be called an innovation (Garcia & Calantone, 2002). Patents, on the other hand, safeguard the rights of inventors whose inventions are really breakthrough and economically successful by assuring that an inventor has control over the commercial use of their creation (wipo.int, 2023). As a result, because innovations are the foundation of innovation and represent a novel solution to a technological challenge, they can be protected by patents (wipo.int, 2023). Accordingly, in 1975, a standard patent categorization was established for ideas that included published patent applications, inventors' certificates, utility models, and utility certificates (WIPO, 2023). This categorization was provided by the Strasbourg Agreement on International Patent Classification (IPC) (WIPO, 2023). Patent information can be searched through online resources that incorporate patent databases (Singh et al., 2016). These databases contain both structured and unstructured data related to the inventor, patent attorney, citations, patent names, summaries, claims, and descriptions (Singh et al., 2016). To search for patent information, various techniques, such as data mining or text mining, are commonly used (Singh et al., 2016).

HRM is the efficient and effective use of human resources to fulfil an organization's goals (Opatha, 2009). According to Armstrong (2006), HRM pertains to the consistent administration of a company's valuable resources, therefore, its employees. These individuals, both, individually and collectively, play a role in achieving the organization's goals. This field encompasses procedures such as hiring, choosing candidates, providing training, fostering development, and overseeing performance to ensure that the workforce aligns with the organization's overall strategic vision (Armstrong, 2006). HRM is also known as Personnel Management, Manpower Management, People Management, and Staff Management (Opatha, 2009).

HRM innovation refers to innovative ideas that are integrated into a company's human resources programs, processes, and practices to provide direct or indirect value to the organization (Amarakoon et al., 2018; Wolfe, 1995; Wolfe et al., 2006). According to Birkinshaw et al. (2008), HRM innovations fall under the category of non-technical managing innovations, which are defined as novel management practices, processes, structures, or techniques that are intended to further organizational goals (Amarakoon et al., 2018). However, the emergence of various technologies, particularly of the World Wide Web, has revolutionized human resources processes and employment connections (Kiesler et al., 1984; Stone & Deadrick, 2015). Information technology (IT) was especially crucial in the era of the COVID-19 pandemic, when several changes were introduced in HRM, such as virtual meetings through various internet platforms, and virtual training for employees (Azizi et al., 2021). Also, technological innovativeness used in HRM are advanced technologies, artificial intelligence technologies, and robotic technology (Vrontis et al., 2022).

Having in mind all of the above, the goal of this chapter is to analyze and get insights into the innovations related to the technical field of HRM. The chapter will provide answers to the following research questions: (RQ1) What is time wise publication trend in innovation activities related to HRM?; (RQ2) Who are the top assignees and inventors within the technical field related to human resources management?; (RQ3) How are innovation activities related to the HRM spread across different countries?; and (RQ4) What is the technical content of HRM innovations?. In order to answer these questions, the authors performed a patent search using PatSeer databases. Selected data consisting of 2109 patent families regarding HRM was analyzed through the patent landscape and text mining approach. For this purpose, MS Excel and Provalis Wordstat have been used.

Aiming to fulfill the stated goal of the paper and to answer the research questions, the structure of this chapter is as follows. After the introduction, the second section presents the background of HRM innovations, previous research regarding patent analysis for the technical field of HRM, as well as covers the methods that were used in previous research for patent analysis. The third section describes the methodology of this research. The results are presented in the fourth section, following with the discussion. Finally, the conclusion gives the findings, limitations, and future research directions.

2 Theoretical Background

This part of the paper provides theoretical background of the HRM field, with a special focus on innovation. In that sense, the innovations for HRM are discussed, as well as the approaches and methods of the patent analysis, following with a brief overview of the research conducted in the patent analysis for the technical HRM field.

2.1 Innovations for Human Resources Management

The term "innovation" can have several meanings and so can be defined in various ways. Everett Rogers, the author and creator of the "Diffusion of Innovations" theory and model, defined innovation in 1995 as "any new thought, thing, or behavior that is not usual for an individual or, in this case, an organization". According to Porter (1990), any innovation is a process in which knowledge and technology are employed to develop new or modified/improved current products of the business. In the context of the process, innovation does not always reflect merely the output of a new product, service, or technology, but also a targeted change in an organization's social or economic potential (Baars & Kemper, 2008). Innovation, in addition to being seen as a process, may also be viewed as a new service, product, or style of organization that can alter the essential functions of the business (Ackah, 2017). In that sense, De Leede and Looise (2005) list the following crucial aspects of innovation: (i) change that results from the introduction of anything novel, at least for the current organization, such as novel goods or services, novel technology, or novel organizational structures; (ii) having a process dimension, which includes actions/phases including planning, organizing, and carrying out the goals; (iii) a change characterized by noticeable discontinuities, notwithstanding the fact that many writers often refer to incremental innovation (or continuous innovation); and (iv) the goal of improving the company's position.

In the 1990s, the IT innovation market remained one of the fastest-growing innovation markets (Freeman & Perez, 1988; King et al., 1994; National Research Council, 1988, 1994; Willinger & Zuscovitch, 1988), and future growth is anticipated. According to Business Research Company (2023), the total value of the worldwide IT market increased from 8179.48 billion US dollars in 2022 to 8852.41 billion US dollars in 2023, representing a compound annual growth rate (CAGR) of 8.2%, and at a CAGR of 7.9% to 2027, the IT market is forecast to be worth 11,995.97 billion US dollars. Numerous researches showed that IT has a strong impact on innovation, which also impacts organizational performance. For example, Naidoo and Hoque (2018) investigated the strategic relevance of IT by examining its role in the relationship between innovation capability and firm performance and revealed a strong positive relationship between innovation and firm performance, as well as a moderate relationship between IT capability and innovation capability. Alam et al. (2019) examined the impact of IT on innovation capability, competitiveness, and organizational performance and reported that the use of IT has a direct and positive effect on innovation capability, competitiveness, and organizational performance and that innovation capability also has a direct and positive effect on competitiveness and organizational performance. Given that IT innovations have a significant impact on organizational change, as well as organizational innovation in terms of its creativity and flexibility in business (Dougherty, 1992; Henderson & Clark, 1990; Le Bas et al., 2011), it is critical to constantly monitor and keep up with them.

According to Singh and Kumar (2022), innovations in HRM include the adoption of technology to streamline processes and reduce administrative burden, improve access to data for employees and managers, provide real-time metrics for decision-making, and enable human resources to play a more strategic role in the organization. They also stress out that human resources are evolving into a more technology-based profession (Singh & Kumar, 2022). Innovation capacity is seen as essential for organizations to respond to changing environmental conditions and deliver better performance (Dede, 2019). The complex and dynamic nature of HRM and development, learning, knowledge management, and innovation require a multilevel consideration and a holistic theoretical approach (Loewenberger, 2016). Also, in the knowledge economy era, the innovation of HRM is crucial for enterprises to form high-quality teams and constantly update products and services (Sun, 2013). In that sense, Machado and Davim (2014) point out innovative climate, alongside flexibility, collaborative working environment, informal communication and knowledge workers as one of the strengths of HRM in technology start-ups and explain that knowledge workers consistently strive for novel ideas and concepts. Additionally, they also argue that a culture that acknowledges failures as an integral component of the innovation process also fosters the advancement of new product development (Machado & Davim, 2014). In that context, while investigating innovations in HRM, Sajjad et al. (2020) focus on addressing workplace loneliness and its impact on creativity, and identify intrinsic motivation as a key mediator between loneliness and employee creativity.

2.2 *Patent Analysis—Approaches and Methods*

The terms "innovation" and "technological advancement" have taken on increasingly important roles in the modern context of HRM. Patents, which are legal mechanisms that protect unique ideas, serve a crucial role in ensuring the continued success of these breakthroughs while also preventing their theft. Within the realm of technical HRM, this chapter goes into the difficult process of patent analysis and what it entails.

It provides insights into the approaches and concepts that are required for a full grasp of the patent environment pertaining to this discipline. In that sense, it is important to define patents and explain the concept of patent analysis.

Rosenberg (1992) defines patents as a legal agreement between an inventor and the government that grants the inventor the right to exclude others from making, using, or selling the invention for a limited period of time. Emma (2005) adds that patents are legal documents that clearly articulate the specific invention being protected, while Li and Meyer (2014) define patent as a kind of intellectual property protection that provides the owner with the ability to exercise control over how their innovation is used and how it is commercialized. Also, White (2009) describes patents as essentially restricted monopoly rights that are awarded by governments. Accordingly, every inventor has a legal right to be exempt from the use and production of a certain invention, and this right is called a patent (Hall, 2007). Such a legal right encourages new innovations and at the same time protects the intellectual rights of every inventor (Hall, 2007). It is the patent law that encourages inventors to apply for patents early in the process of creating innovations before their production and following commercialization (Sichelman, 2009). Hamilton and Till (1948) explain that when an invention is submitted for patent protection, it must demonstrate both originality and inventiveness in order to be protected. Tyczewska (2014) explain how patents encourage innovation by giving inventors exclusive rights to their inventions for a limited period of time, serving as an incentive for inventors to create new technologies. Several other authors also agree and emphasize that patents encourage further development and inventions and also reward scientific endeavors (e.g. Li & Meyer, 2014; Rosenberg, 1992). In addition, the use of patents serves to enhance the dissemination of valuable information by necessitating innovators to publicly reveal their innovations., which helps in advancing the quality of life (Elliot, 2007), and they also provide legal safeguards in product markets, protecting the rights of inventors and allowing them to exclude others from practicing their inventions (Kultti et al., 2007). Hsu and Ziedonis (2013) further argue that patents can also serve as quality signals, improving access and the terms of trade in factor input markets, while Sadiku (1997) explains that a patent system makes collusion among innovators more difficult, reducing spillovers and eliminating collusion. Hall (2007) discusses the effectiveness of patents in encouraging innovation and reviews controversies in patent policy and concludes that the overall policy attitude of many countries is to encourage businesses, people, and research organizations to learn about the patent system and file for patents. Nevertheless, Walker (1995) highlights the value of patents as a source of scientific information and describes various methods for accessing patent information. In accordance, White (2009) also sees patent documents as the valuable sources of scientific information. Having in mind the previous, one can conclude that patents can be seen as legal agreements that grant inventors the right to exclude others from using their inventions, and can be valuable sources of scientific information.

Patent analysis is a research method that involves analyzing, describing, and interpreting patents in a systematic way. According to Park and Jun (2022), patent analysis is a method used to analyze patent data in order to understand a specific technology or area of innovation, while Cruz et al. (2022) see it as a tool designed to transform statistical data related to patents into relevant and practical information tailored to meet particular requirements. It involves studying the information present within and attached to patents, such as the title, abstract, claims, inventors, applicants, and jurisdictions (Fatimi, 2022).

There are many areas of application where patent analysis can be useful. In that sense, Wilson (1987) discusses the potential of patent analysis as an econometric measure and a valuable aid to decision-making in areas such as acquisitions, planning of research and development, and new product development. In accordance, many authors also see patent analysis as a valuable tool for strategic decision-making, as it can inform organizations about emerging technologies and guide their research and development efforts (e.g. Daim et al., 2006; Harhoff & Wagner, 2009; Lee et al., 2008, 2009; Trippe, 2015). Another patent analysis area of application is given by Phadnis and Hirwani (2005) who demonstrate how patent analysis can be used as a tool for research planning, using the example of analyzing patents related to tea's medicinal properties. Furthermore, Aithal and Aithal (2018) explain that patent analysis is a qualitative research method whose main goal is to evaluate patent's merits, drawbacks, limitations, efficacy and potential future value, from which a theory or a new concept can be developed.

Patent analysis can provide insights into the state of innovation in a particular field, including the number of patent applications, the leading countries and institutions, and the focus of research and development (Aithal & Aithal, 2018; Zhou & Wang, 2021). It can also help identify technological trends, potential applications, and areas for further research (Choi & Song, 2018; Elvers et al., 2016; Kwon et al., 2022). In that context, patent analysis is a useful research method for identifying trends, research gaps, and opportunities in various fields, and, at the same time, ensuring originality of work and informing research and innovation funding policy (Klongthong et al., 2021; van Rijn & Timmis, 2023).

According to Aithal and Aithal (2018), there are two distinct categories of patent analysis: (i) individual patent analysis, and (ii) group patent analysis. The process of individual patent analysis involves the examination of a patent that has been submitted by a person, organization, or corporation for a product, technique, or technology, while group patent analysis involves the comparison of a collection of patents that are connected either by industry or by the same technology (Aithal & Aithal, 2018). Aithal and Aithal (2018) also explain that multiple analytical frameworks may be used to perform patent analysis, including patent opportunity analysis, patent performance analysis, patent innovation analysis, patent technology analysis, and patent value analysis.

Patent landscape is a method used in patent analysis to understand the trends and patterns in a specific technological domain (Trippe, 2015). To put it in other words, patent landscape refers to the process of analyzing patents related to a particular topic. Yang et al. (2010) define patent landscape as "a state-of-the-art patent search that provides graphic representations of information from search results", while Adel (2021) defines it as "a portfolio of patent publications or patent families which meet a specific search criterion from among the approximately 14 million patents currently in force and the many more which have lapsed or are pending". Yang et al. (2010)

continue to explain that patent landscape analysis report commonly includes three main components: (i) a visual depiction of information obtained from a comprehensive search of existing patents; (ii) graphics and charts that illustrate patterns in patenting activities, identify prominent patent holders, highlight collaborative partnerships, and evaluate technological advancements; and (iii) textual information that corresponds to the graphs and charts within an interactive database accompanying the analysis report. Its goal is to provide clear and actionable answers to particular policy or practical concerns (Trippe, 2015). Historically, industries have relied on patent landscapes to inform their choices on capital expenditures, research and development priorities, competitive threats, and product launch flexibility (Trippe, 2015). Trippe (2015) indicates that, in recent times, there has been an increasing use of patent landscapes by public policymakers as a means to establish an empirical basis prior to deliberating on overarching policy issues, which is notably evident in domains such as health, food security, and the environment. Similarly, Soni et al. (2013) explain that patent landscaping is a comprehensive analysis report of patents, scientific literature, market research, and survey data, which elaborates the white space in the research area and intellectual property trends, and enables strategists, scientists, and attorneys to take key decisions in new product development, research & development planning, and strategic development. In that sense, many authors use patent landscape analysis in many different research areas. For example, Aggarwal and Chandra (2021) provide an example of a patent landscape report on plant stem cell-based patents. Androshchuk and Kvasha (2019) present a method of using patent analysis to determine perspective directions of technological development in the field of armaments and military equipment, while Abood and Feltenberger (2018) present an automated approach to patent landscaping that leverages human domain expertise, heuristics based on patent metadata, and machine learning to generate high-quality patent landscapes with minimal effort. Similarly, Forestal (2023) preforms an examination of the biopiracy of indigenous plants throughout the island of Hispaniola by the use of an automated patent landscaping and legal geography analysis. Patent landscaping has been used also by Nomaler and Verspagen (2021) to identify the main technological trends in green technology and to map out specific green technologies and their relationships with non-green technologies.

Mao et al. (2022) argue that traditional methods for technology analysis were qualitative, but recent approaches have introduced statistical and machine learning techniques. In that sense, Tang et al. (2012) argue that there are many challenges related to performing a patent analysis in the twenty-first century as there is a huge volume of patents that need to be searched in order to find relevant patents and related information. Therefore, Tang et al. (2012) present a topic-driven patent analysis and mining system that can be used for intellectual property and research and development strategy planning, while Tseng et al. (2007) describe a series of text mining techniques that can assist patent analysts in analyzing patent documents, including text segmentation, summary extraction, feature selection, term association, cluster generation, topic identification, and information mapping. In that sense, in order to facilitate the research process giving the huge amount of patent data, many researchers use text mining techniques for patent analysis. For example, Wich et al. (2013) demonstrate the use of a text mining solution for the analysis of patents related to the development of a new method for repairing gas turbines, while Kim et al. (2019) conduct a patent analysis using text mining techniques to uncover developing and unexplored technical domains within the field of wireless power transfer. Bamakan et al. (2021) examined the blockchain technology patents using text mining and clustering techniques and predicted blockchain technology trends based on its classification by the World Intellectual Property Organization (WIPO) database. All of the examples mentioned above, alongside the many more published in the literature, speak in favor of growing popularity of patent analysis and the usage of methods and techniques such as patent landscaping and text mining.

2.3 Patent Analysis for Technical Field of Human Resources Management

Over the years, in addition to globalization, economic trends, trends in talent diversity and business strategies, the practices and methods of human resources management have also been influenced by technological trends (Dutta, 2018). Accordingly, the topic of technological trends that are protected by patents in the field of HRM has become a current topic in scientific circles (e.g. Dutta, 2018; Mushkudiani et al., 2020). The prompt development of information and communication technology also adds to this, with technology penetrating all workplaces at a pace earlier inconceivable (Przytuła et al., 2020). Namely, the digital transformation has ensured that a large number of employees accept and regularly use digital and social media for the purpose of communication and mutual cooperation (Dutta, 2018).

Consequently, by collecting data about employees that have become available on social media, human resources can use such information in their business processes and thus adjust their methods and practices (Dutta, 2018). Therefore, human resource departments' styles and procedures are continuously evolving (Mushkudiani et al., 2020) and simultaneously modifying with IT development. In the same way, the COVID-19 pandemic and remote working contributed to the use of various advanced technologies in the human resources management process during the process of selecting candidates, their employment, and evaluating their work (Przytuła et al., 2020). According to Mushkudiani et al. (2020), technological trends that had an impact on HRM were mainly related to social networks such as Twitter, Linkedin and Facebook, mobile applications that facilitated tracking the location of employees, websites that allowed employees to access gamification features related to various human resource management processes, cloud computing and data analytics that facilitated the monitoring and evaluation of teamwork, etc.

Accordingly, many authors conduct the analysis of patents from the technical field in HRM processes in order to determine the level of technology used or to improve the strategy of technology development through the analysis of patent data that includes various information about technologies (Chun et al., 2021). Namely, as mentioned before, patent documents are very important for research and development because they contain valuable information such as summaries, titles, and IPC classification codes for specific technical areas (Chun et al., 2021).

According to Vlad (2010), different authors of scientific papers use information about patents in various fields, including technology estimation and HRM. Also, according to Ejermo and Karlsson (2006), patent databases that can be used for various analyses for technology forecasting, social network analysis, and co-inventor network analysis are patent databases such as the USPTO, Derwent Innovation and Google Patents, as well as many others (Zhu et al., 2019).

Accordingly, Kim and Bae (2017) conducted research using patent analysis to predict future promising technologies. Furthermore, Azis et al. (2022) investigate technological trends in the field of composting by using the technology patent landscape for the selected field and analyzing a total of 457 selected patents. Similarly, Maghsoudi et al. (2023) used patent analysis in their study to predict Internet of Things technology trends and to find the main people and companies registering patents in this technological area, all with the purpose of creating a network of collaborators for this observed technology.

On the other hand, Pawar and Shaikh (2021) conduct two case studies in their research to show how patent literature analysis can be an effective tool in the process of hiring human resources in various technological fields. In addition, in their study, Chung et al. (2021) conducted an inventor profile mining approach that was based on patents, all for the ultimate purpose of identifying potential human resources, and what they really wanted to find through patent analysis were the inventors. The background of the research was focused on technology-oriented companies that need employees with high technical knowledge and skills in order to be able to successfully and efficiently implement the technological strategies of each individual company (Chung et al., 2021). In a similar way, Moehrle et al. (2005) use semantic analysis of patents and multidimensional scaling in order to find appropriate profiles and technological competencies of inventors, which are necessary for various decisions in the management and development of human resources of companies. Furthermore, Zhu et al. (2019) use the patent data mining method in order to recognize the so-called "rising technological stars", i.e. the profiles of inventors from technical fields for companies that are looking for a workforce that will contribute to technological innovation, vitality and teamwork in the company and that will meet the needs of the individual, group, department and company. Furthermore, Fareri et al. (2023) conducted an analysis of patents in the field of green computing innovations in order to establish a tool that will detect the lack of digital and green skills in the profiles of employees from the field of energy.

To the best of the authors' knowledge, although there are studies that use different methods in the analysis of patents from the technological field to support HRM, a lack of research dealing with comprehensive technological trends in HRM has been noticed. Research mostly focuses on the analysis of patents that touch on certain tasks in the HRM process, such as technology forecasting, profiling of potential employees for a particular company, finding specific technological knowledge and skills, or profiles of inventors that will support the innovation of the company's operations. For this very reason, the contribution of this chapter is the expansion of the literature from the selected field, by gaining a broader insight into the directions of technological trends that are currently used in HRM processes using the patent landscape and the analysis of the technical content of patents.

3 Methodology of the Human Resources Management Innovation Analysis

This section provides an overview of the methodology employed to perform HRM innovation analysis. For this study, a three step methodology has been used. The first step included searching and selecting data. The second step included the creation of patent landscape and the third step refers to the analysis of technical content of patents. All three steps are described in more detail in the continuation.

3.1 Step One: Searching and Selecting Data

Prior patent search authors prepared a combination of search terms. The aim was to collect data for patents related to human resources management. However, it was necessary to assume that patents may be introduced through different synonyms. The term "human resources management" can be substituted by similar phrases which should not be ignored. Therefore, a synonym list was created with some help from an English language expert, thesaurus and semantic recommender to identify common English language synonyms for the term "human resources management": Additionally, spelling has been taken into account, e.g., labor and labour. Several combinations of search string have been used to search the PatSeer database on 17 August 2023. Optimal results have been achieved by using search string T: ("human resource management" OR "human resources management" OR "management of human resources" OR "personnel management" OR "people management" OR "staff management" OR "workforce management" OR "labor resource management" OR "labour resource management" OR "labour resources management" OR "labor resources management" OR "hr management" OR "management of human resources" OR "personnel management" OR "administration of human resources" OR "HRM" OR "human capital management" OR "human-resources management" OR "human-resource management" OR "human resources administration") with an option for searching patent families. Hence, records which had any of the keywords in their titles have been selected. It is also important to explain that single innovation can be protected in various countries which makes a patent family. For example, if a company files a patent application in China and then extends it to the USA, then the USA application is a family member of the China application (Sinha & Pandurangi, 2016). For that reason, patent families instead of patents have been searched.

Authors extracted 2237 records for patent families in total. Among these records, 2109 contained technical content related to HRM. The legal status of a patent indicates whether a patent is active (granted or applied) or inactive (withdrawn, abandoned, or expired). It is "data related to events in the lifetime of a patent" (Sinha & Pandurangi, 2016, p. 102). Both "active" and "inactive" were used for further innovation analysis. Date restriction was not set while searching for the data but it was restricted for the last five to ten years to derive relevant insights when it comes to the recent publication and content trends.

3.2 Step Two: Patent Landscape

The authors analyzed the timeline, geographic origin, current assignees and inventors to detect which of them were most active in innovating technical content related to HRM. Timeline is analyzed by using data on publication year which indicates the date on which the patent is published by the respective patent office. Furthermore, the authors carried out an analysis of countries in which HRM innovations were published. Assignee is the name of an individual or company applying for a patent and the current assignee is the latest assignee of a patent document (Sinha & Pandurangi, 2016). "Inventor is a person or persons who have invented a technology" (Sinha & Pandurangi, 2016, p. 102). Microsoft Excel was used in this step.

3.3 Step Three: Analysis of Technical Content of Patents

The authors analyzed the technical content of HRM patent families according to the International Patent Classification (IPC) system. In this research, the analysis of HRM patent families according to the sections, main and full content have been conducted.

To explore the content of HRM innovations authors have analyzed HRM patent family abstracts by using a text-mining approach. Authors detected the phrases emerging most often in abstracts of patent families related to HRM. They have used WordStat Provalis for that purpose. Before performing a frequency analysis on the phrases encountered in the abstracts, text preprocessing has been conducted.

An important part of text preprocessing was stemming, lemmatization, as well as the creation and management of an exclusion list. Stemming, as a natural language processing routine "reduces inflected and derived forms of words to a common root form or word stem" (Provalis Research, 2021, p. 7). Lemmatization is a process by which various forms of words are reduced to a limited number of canonical forms, for example, lemmatization includes the conversion of plurals to singulars and past tense verbs to present tense verbs (Provalis Research, 2021, p. 7). Stemming and lemmatization for English were used for this study since the patents' abstracts are written in English. Exclusion or stop list contains words and phrases that are excluded from the frequency analysis. It is used mainly to remove words with little semantic value, such as pronouns and conjunctions (Provalis Research, 2021). However, it may also be used to remove phrases that don't bring much novelty to the results of the analysis. Therefore, keywords that were used in the string for patent search in the PatSeer database were excluded from the analysis to detect more interesting HRM innovations content and connections between phrases.

First, phrases of a minimum of three and a maximum of four words that occur in more than ten simple patent abstracts are extracted. Next, for the purpose of this research, the authors have restricted the definition of cooccurrence to entries that appear in the same paragraphs of the abstracts. After counting the frequency of a phrase, the next step is to modify the frequency with the perceived importance of that phrase. The Term Frequency- Inverse Document Frequency (TF-IDF) was used to compute scores for phrases in a collection of abstracts of all HRM patent families. The reason for using the TF-IDF metric is that common words usually appear several times in an abstract, but they are not as important as key phrases.

TF-IDF measure was described in Silge and Robinson (2017). It is the frequency of a phrase adjusted for how rarely it is used. Term Frequency (TF) indicates how frequently a phrase occurs in an abstract of the patent family. Inverse Document Frequency (IDF) measures the importance of a phrase concerning the whole collection of patents' abstracts. The IDF for a given phrase in the collection of patent families is calculated as:

$$IDF = lnln\left(\frac{n_{abstracts}}{n_{abstract\ containing\ phrase}}\right)$$

The product of TF and IDF value gives TF-IDF value for a phrase. The higher TF-IDF value for a phrase is more important a phrase is.

Second, extracted phrases were used to conduct cluster analysis to detect which phrases occur together. Cluster analysis of phrases was conducted using hierarchical clustering algorithm. The similarity measure used in clustering and multidimensional scaling was association strength. It is a probabilistic measure that measures the number of expected co-occurrences for two phrases assuming the possibility that two phrases will sometimes co-occur by chance (Steijn, 2021). Results are displayed in the form of dendrograms, and circular graphs based on phrase co-occurrence. In the dendrogram the vertical axis is made up of the phrases and the horizontal axis represents the clusters formed at each step of the clustering procedure. Phrases that tend to appear together are combined at an early stage while those that are independent from one another or those that don't appear together tend to be combined at the end of the agglomeration process. Circular graphs were used to visualize the connections between phrases and to explore relationships as well as detect underlying patterns.

4 Results and Discussion

The section describes the results of searching and selecting data, patent landscape and analysis of the technical content of HRM patents. Also, the discussion of the results is provided.

4.1 Step One Results: Patent Families Sample

After data preparation, the authors decided to analyze 2109 records that contained technical content related to human resource management (HRM). Figure 1 provides details on the legal status of selected HRM patent families. It can be noticed that 57% of HRM patent families are active (applied or granted) and almost 43% are inactive due to withdrawal, rejection, nonpayment or expiration.

Each of the HRM patent families was registered to one or more than one IPC's sections, main groups or sub-groups. Therefore, the total number of ICR codes is usually larger than the number of patent families. Table 1 presents the number of simple patent families according to IPC section.

According to Table 1, HRM patent families were registered under eight IPC sections. Most of them (56%) are part of the IPC technical section G: Physics. Significant number of patent families belong to section A Human necessities (16% of patent families that belong, section B: Performing operations; Transporting (11%

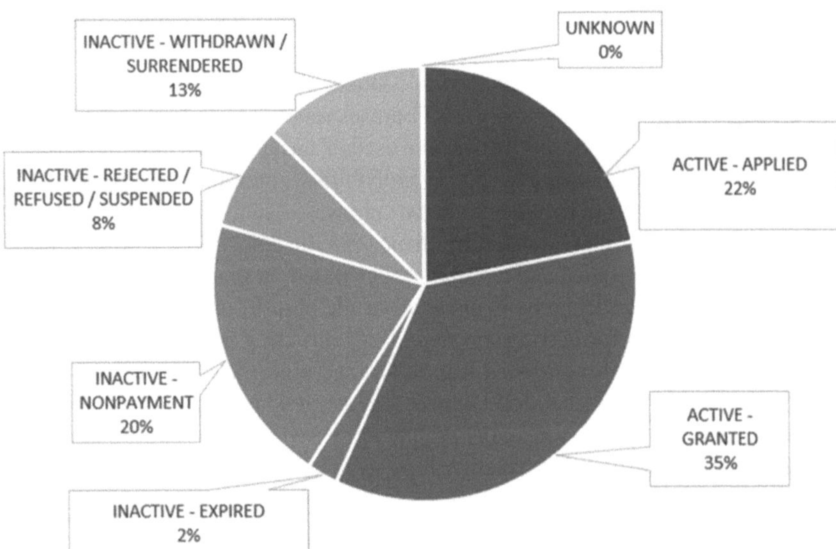

Fig. 1 Current legal status of sample HRM patent families (*Source* Authors' work, 2023)

Table 1 Number of simple patent families according to IPC section (*Source* Authors' work, 2023)

IPC section	Total number of patent families	% of total number of patent families
G: Physics	1589	56
A: Human necessities	437	16
B: Performing operations; transporting	303	11
H: Electricity	238	8
F: Mechanical engineering; lighting; heating; weapons; blasting	156	6
E: Fixed constructions	43	2
Unknown	31	1
C: Chemistry; metallurgy	20	1
D: Textiles; paper	2	0

of patent families), section H: Electricity (8% of patent families) and F: Mechanical engineering (6%). Some of HRM simple patent families are also assigned to sections E: Fixed constructions, C: Chemistry; Metallurgy and D: Textiles; Paper. Giving that HRM patent families span eight distinct IPC sections, it could be concluded that HRM is an interdisciplinary field. Such variety reflects the multifaceted nature of HRM, which, according to the results, includes everything from the management and organization of administrative tasks (physics and human needs) to the development and implementation of technological solutions (electrical and mechanical engineering).

4.2 *Step Two Results: Patent Landscape*

Patent landscape results contain an overview of the HRM technical field providing multiple insights for understanding the field.

Table 2 provides details about the geographical distribution of patents specific to HRM. China's assignees lead in patenting activities in the technical field of human resources management with almost 77% of the total number of HRM patent families. Other assignees are spread across India, Japan, the United States of America and South Korea. On the other hand, Spain, Romania, Ukraine and Singapore each have only one patent families noted in this study, which counts for 0.05% of the total number of patent families for each named country. One should also note that there are five patent families whose publication country is unknown, which accounts for 0.24% of the total number of observed patent families.

Table 3 describes the publication timeline for HRM patents across the top five countries—China, India, Japan, South Korea and the United States of America for

Table 2 Number of HRM patent families according to publication country (*Source* Authors' work)

Publication country	Number of patent families	% of total number of patent families
China (CN)	1620	76.81
India (IN)	146	6.92
Japan (JP)	97	4.0
United States of America (US)	97	4.60
South Korea (KR)	49	2.32
Australia (AU)	36	1.71
Taiwan (TW)	26	1.23
Canada (CA)	9	0.43
Germany (DE)	6	0.28
Unknown	5	0.24
Turkey (TR)	4	0.19
Russia (RU)	4	0.19
United Kingdom (GB)	2	0.09
World Intellectual Property Organization (WIPO)	2	0.09
South Africa (ZA)	2	0.09
Spain (ES)	1	0.05
Romania (RO)	1	0.05
Ukraine (UA)	1	0.05
Singapore (SG)	1	0.05
Total	2109	100

the period between 2012 and August 17th 2023. It is noticeable that the highest number of HRM patents for these countries were published in recent years, i.e. 2021 and 2022. In the current year, 2023, and among the five observed countries, the most HRM patents has been published in China (93 of them) and the least of them in Japan (3 of them). These results indicate a recent surge in HRM innovation and technological advancement which may be driven by the need for more efficient human resources processes, especially in response to challenges posed by the recent COVID-19 pandemic.

Figure 2 shows the timeline for the total number of HRM patent families for the period between 2012 and August 17, 2023. As it can be seen from the Fig. 2, there is a significant growth in the total number of HRM patent families from 2017 which peaks in 2021 with 481 HRM patent families. In 2022, there is a decrease in patenting activities. According to the current number of HRM patent families, a further decrease can be foreseen for 2023 as well.

Table 3 Number of HRM patent families according to top five publication countries and publication year (*Source* Authors' work)

Publication year	China	India	Japan	South Korea	United States of America
2012	16	1	5	2	3
2013	13		3		9
2014	24		3		7
2015	41	1	1	2	2
2016	37	1	3	2	9
2017	53		1	1	5
2018	89	1	2	3	4
2019	155	1	1	2	6
2020	323	2	8	3	4
2021	423	15	17	9	8
2022	327	76	8	7	6
August 17th 2023	93	46	3	8	4

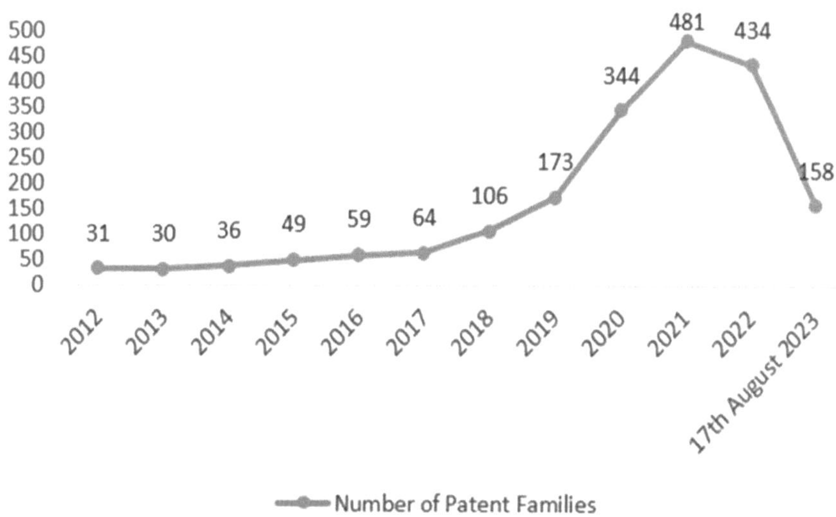

Fig. 2 Number of HRM patent families according to publication year from 2012 to 17th August 2023 (*Source* Authors' work, PatSeer and MS Excel, 2023)

Tables 4 and 5 show key players who are working in the technological field of HRM. Most of the key players are from China and Japan. The results also indicate that key players usually come from business or academia.

When it comes to technological wise trends, the number of patent families can be observed within technological subdomains. Figure 3 shows technical subdomains with more than 30 patent families for the period from 2019 to 17th August 2023. A

Table 4 Top current assignees according to number of HRM patent families (from 2012 to 17th August 2023; ≥ 5 patent families). *Source* Authors, PatSeer and MS Excel, 17th August 2023 (*Source* Authors' work, 2023)

Current assignee	Number of simple families
Shaanxi University of Science and Technology (China)	11
Hangzhou Yazhang Tech Co Ltd. (China)	11
DISCO Corporation (Japan)	11
SAP SE (Germany)	8
Jiangsu Vocational Institute of Commerce (China)	8
Gansu Zhongda Human Resources Service Co Ltd (China)	8
Toshiba Digital Solutions Corporation, Toshiba Corporation (Japan)	6
Shenyang University of Chemical Technology (China)	6
The Second Affiliated Hospital of Zhengzhou University (China)	6
Jiaozuo University (China)	5
Guizhou Vocational Technology College of Electronic & Information (China)	5

Table 5 Top inventors according to number of HRM patent families (from 2012 to 17th August 2023; ≥ 5 patent families). *Source* Authors, PatSeer and MS Excel, 17th August 2023 (*Source* Authors' work, 2023)

Inventor	Number of simple families
Li Haobin	12
Sekiya Kazuma	11
Feng Hujun	9
Liu Qinyi	5

single patent family can be filled in more than one technical subdomain. For example, patent family x can be filled as an IT method for management and as computer technology. The highest number of patent families have been filled in the following technical subdomains: IT methods for management; furniture, games; control, audio–visual technology; both as IT methods for management and as computer technology; other consumer goods; computer technology and both as computer technology and IT methods for management.

Again, the results presented by the Fig. 3 show the diversity and convergence of technologies in HRM, as well as the potential for enhancing the employee experience, data-driven HR practices, and the need for organizations to adapt to evolving work environments. Leveraging these technological innovations effectively can position organizations for success in talent management and workforce optimization.

Next, Fig. 4 shows how HRM patent families have been evolving over the years in technological subdomains with more than 30 patent families for the period from 2019 to 17th August 2023. The number of patent families grows and peaks in 2021 for

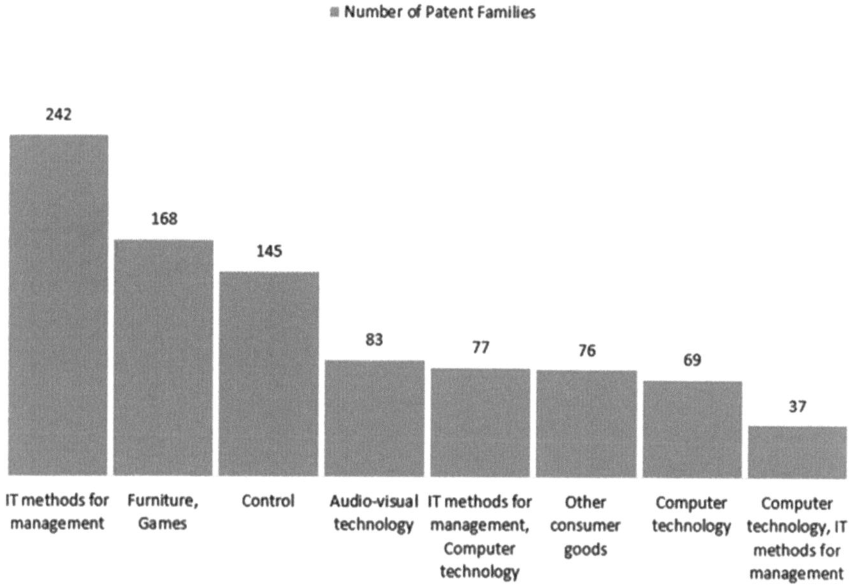

Fig. 3 Number of HRM patent families according to technological subdomains with more than 30 patent families (from 2019 to 17th August 2023) (*Source* Authors, PatSeer and MS Excel, 17th August 2023)

the technical subdomain of furniture, games; control, and audio–visual technology; both as IT methods for management and as computer technology; other consumer goods; computer technology and both as computer technology and IT methods for management. The number of patent families in IT methods for management has decreased from 2019 to 2020 and after that, it grows until 2022.

Figure 5 describes the number of HRM simple patent families for the top five publication countries according to technological subdomains for the period from 2019 to 17 August 2023.

As it can be seen from the Fig. 5, China has published HRM patent families for all of the technical subdomains, but most of them are related to furniture, games, control and IT methods for management. Assignees from India usually have been published patent families for IT methods for management and control. Japanese, South Korean and US assignees are focused on IT methods for management.

Since China's assignees have published most of the simple patent families, it was decided to explore their innovation activities according to technological subdomains in more detail. Figure 6 shows that China's assignee patenting activities have been changing for the past three years. In 2019 most of the patents were related to IT methods for management. However, patenting activities regarding furniture and games have significantly increased in 2020 and 2021 compared to 2019 and took first place in 2021.

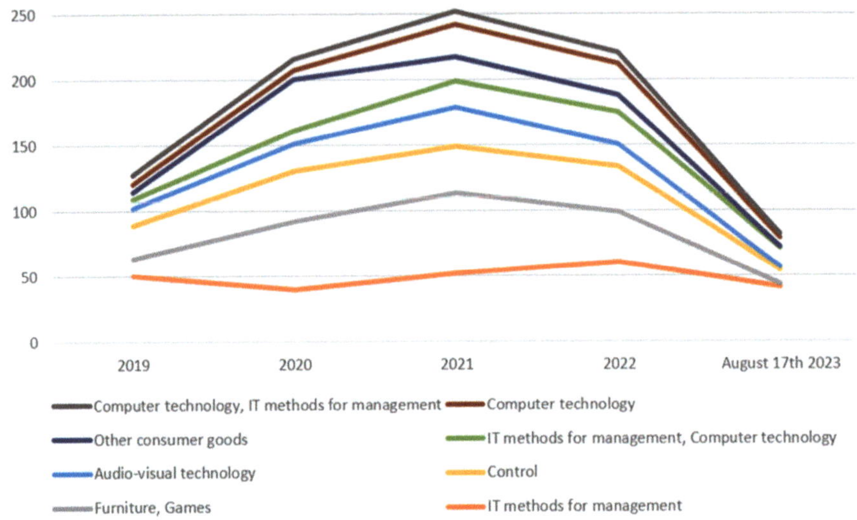

Fig. 4 Number of HRM patent families per year according to technological subdomains with more than 30 patent families (from 2019 to 17th August 2023) (*Source* Authors, PatSeer and MS Excel, 17th August 2023)

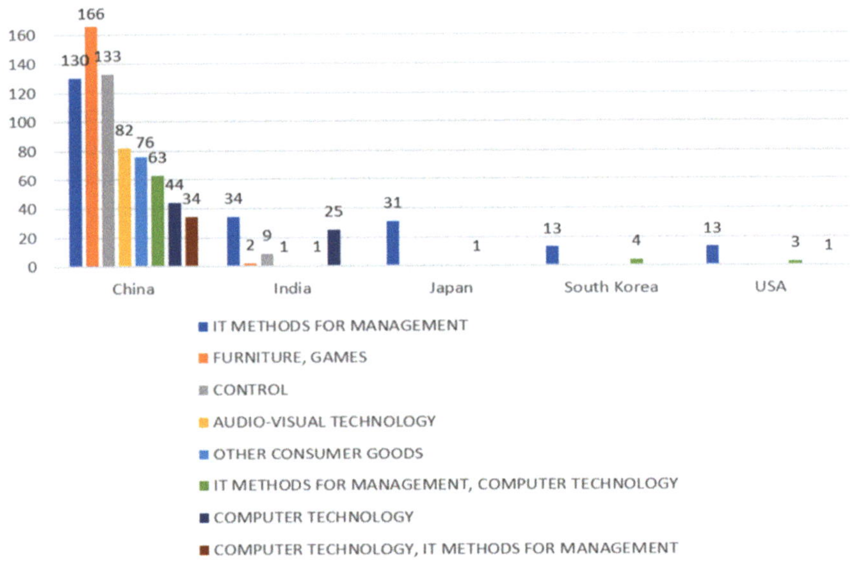

Fig. 5 Number of HRM simple patent families for top five publication countries according to technological subdomains (from 2019 to 17th August 2023). *Source* Authors, PatSeer and MS Excel, 17th August 2023)

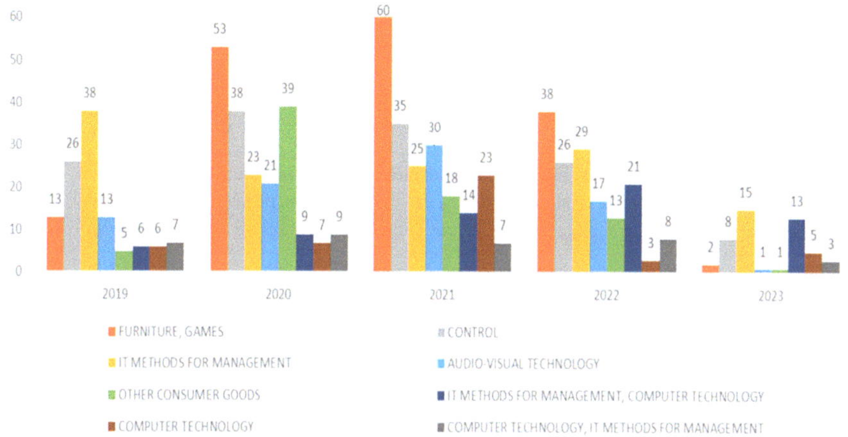

Fig. 6 Number of HRM simple patent families published by China according to technological subdomains with more than 30 patent families publication year (from 2019 to 17th August 2023) (*Source* Authors, PatSeer and MS Excel, 17th August 2023)

4.3 Step Three Results: Extracted Content of Patents

In this step, patents are categorized in a series of technology-based hierarchies under IPC taxonomy. This analysis is also called a domain-focused landscape. After that, abstracts of patent families were analyzed by using text mining to provide details of content and phrases covered by the HRM families.

Table 6 shows what are IPC's main categories that occur in more than 20 HRM patent families and the number of patent families that have been filed in those categories.

As seen in the Table 6, among the main categories assigned to HRM patent families is computing, calculating or counting technical content such as G06Q—Data processing systems or methods, specially adapted for administrative, commercial, financial, managerial, supervisory, or forecasting purposes (907 patent families); G06F—Electrical digital data processing (271 patent families), G06K—Recognition of data; presentation of data; record carriers; handling record carriers (178 patent families). Such high number of patent families in the named three categories suggests that there have been substantial technological advancements in HRM, and the emphasis on data processing and digital data handling (G06F and G06K) implies that HRM has become increasingly automated and data-driven. Also, technologies related to data processing and management (G06Q) can provide human resources professionals with enhanced decision support tools and can assist in strategic human resources planning, resource allocation, and forecasting.

Furthermore, HRM patent families include checking devices such as that were G07C—Time or attendance registers. Also, there are patent families that are protected as furniture such as A47B: Tables; desks; office furniture; cabinets; and drawers (339 patent families). HRM patent families also protect educational and displaying

Table 6 Number of HRM patent families according to the IPC system—main level (>20 simple patent families) (*Source* Authors' work, 2023)

IPC main	Number of patent families
G06Q: Data processing systems or methods, specially adapted for administrative, commercial, financial, managerial, supervisory, or forecasting purposes	907
G07C: Time or attendance registers	339
A47B: Tables; desks; office furniture; cabinets; drawers	327
G06F: Electric digital data processing	271
G06K: Recognition of data; presentation of data; record carriers; handling record carriers	178
G09B: Educational or demonstration appliances; appliances for teaching, or communicating with, the blind, deaf, or mute; models; planetaria; globes; maps; diagrams	166
G09F: Displaying; advertising; signs; labels or nameplates; seals	125
F16M: Frames, casings, or beds, of engines or other machines or apparatus; stands or supports	107
G08B: Signaling or calling systems; order telegraphs; alarm systems	72
B42F: Sheets temporarily attached together; filing appliances; file cards; indexing	64
G06N: Computer systems based on specific computational models	63
H04L: Transmission of digital information	63
H04W: Wireless communication networks	63
B43L: Articles for writing or drawing upon; accessories for writing or drawing	60
B01D: Separation	54
B08B: Cleaning in general; prevention of fouling in general	54
H04N: Pictorial communication	44
G16H: Healthcare informatics	43
G06V: Image or video recognition or understanding	41
H05K: Printed circuits; casings or constructional details of electric apparatus; manufacture of assemblages of electrical components	38
A61B: Diagnosis; surgery; identification	31
A61L: Methods or apparatus for sterilizing materials or objects	23
B42B: Permanently attaching sheets, quires, or signatures	22
H04M: Telephonic communication	21
E06B: Fixed or movable closures for openings in buildings, vehicles, fences, or like enclosures	20

apparatus and methods such as G09B—Educational or demonstration appliances; appliances for teaching, or communicating with, the blind, deaf, or mute; models; planetaria; globes; maps; diagrams (166 patent families) and G09F: Displaying; advertising; signs; labels or nameplates; seals (125 patent families).

Table 7 presents simple patent families according to IPC full level. The most frequently protected content by HRM patent families is related to resources, workflows, human or project management such as organizing, planning, scheduling, or allocating time, human or machine resources, enterprise planning and organizational models (G06Q10/06, 306 patent families). A significant amount of patent families was related to office automation, e.g., computer-aided management of electronic mail or groupware (G06Q10/10, 278 patent families). Additionally, 200 patent families were registered as cabinets, racks or shelf units, specially adapted for storing books, documents, forms, or the like (A47B63/00). Time or attendance registers together with the recording, indicating, or registering of other data, e.g., of signs of identity (G07C1/10, 177 patent families) also make an important part of HRM patenting activities.

In addition, Table 7 also shows that more than 50 patent families can be noted in three IPC full categories, being the following: (i) boards, hoardings, pillars, or like structures for notices, placards, posters, or the like (G09F15/00, 58 patent families), (ii) administration, e.g., office automation or reservations; Management, e.g. resource or project management (G06Q10/00, 53 patent families), and (iii) teaching not covered by other main groups of this subclass (G09B19/00, 53 patent families).

Overall, the results presented in Table 7 provide insights that can be valuable for understanding the technological landscape and trends in HRM. A strong focus on efficiency, automation, document management, and communication within HRM can be seen. Also, such innovations have the potential to enhance human resources practices, improve employee experiences, and, finally, contribute to organizational success.

Next, authors detected the most frequent phrases in HRM patent family abstracts. The keyword that was used in the search string (human resource, human resource management, personnel management, management system, personnel management system, method and system, human resources, management method and system, human resources management) were excluded from the phrases list to derive more interesting results.

Table 8 shows the most frequent phrases which have a frequency of occurrence $\geq$20. Column TF*IDF contains values that indicate phrases' importance. The most frequent and important phrases are management device, personnel management method, artificial intelligence, human resource management device, human resource management file and file storage.

The dendogram (Fig. 7) presents the results of the cluster analysis that identified six clusters of co-occurring phrases in HRM patent families. The first cluster of phrases includes content related to artificial intelligence, HR management and the Internet of Things. The second cluster contains phrases related to attendance checking devices, attendance machines, card punching, enterprise resources management, construction side and face recognition. The third cluster is related to phrases

Table 7 Number of HRM patent families according to the IPC system—full level (>20 simple patent families) (*Source* Authors' work, 2023)

IPC full	Total number of patent families
G06Q10/06: Resources, workflows, human or project management, enterprise planning; organizational models	306
G06Q10/10: Office automation	278
A47B63/00: Cabinets, racks or shelf units, specially adapted for storing books, documents, forms, or the like	200
G07C1/10: Time or attendance registers together with the recording, indicating, or registering of other data	177
G09F15/00: Boards, hoardings, pillars, or like structures for notices, placards, posters, or the like	58
G06Q10/00: Administration, e.g., office automation or reservations; management, e.g. resource or project management	53
G09B19/00: Teaching not covered by other main groups of this subclass	53
A47B63/06: Office furniture with parts, e.g., trays, movable on pivots	41
G09B5/02: Educational or demonstration appliances with visual presentation; material; to be studied, e.g. using film strip	29
G06K9/00: Methods or arrangements for reading or recognizing printed or written characters or for recognizing patterns, e.g., fingerprints	28
F16M11/04: Means for attachment of apparatus; means allowing adjustment of the apparatus	27
G06F1/16: Constructional details or arrangements	23
B43L1/00: Repeatedly usable boards or tablets for writing or drawing	22
G07C9/00: Individual entry or exit registers	20

blockchain, electronic equipment, human resources management method, personnel management method and storage medium. Furthermore, in the fourth cluster phrases that co-occur are data storage, human resource management data, management device, management equipment and human resource management device. Phrases in the fifth cluster are display board, human resources management teaching, display device and information display. Last, sixth cluster are related to file storage and human resources management file.

Furthermore, circular graphs were generated for the clusters to visualize the connections between phrases and to explore relationships as well as detect underlying patterns and structures. Phrases are represented as nodes. Relationships between phrases are represented as lines connecting those nodes.

Table 8 Most frequent phrases in patent abstracts (≥20 cases) (*Source* Authors' work, 2023)

Most frequent phrases in patent abstracts	Frequency	No. of Cases	% cases (%)	TF • IDF
Management device	70	70	5.8	86.4
Personnel management method	59	59	4.9	77.2
Artificial intelligence	43	43	3.6	62.2
Human resource management device	38	38	3.2	57.0
Human resource management file	35	35	2.9	53.7
File storage	31	31	2.6	49.2
Staff management	25	19	1.6	45.0
Human resource management teaching	24	24	2.0	40.8
Personnel management device	24	24	2.0	40.8
Enterprise human resource management	22	22	1.8	38.2
Information display	21	21	1.8	36.9
File storage device	20	20	1.7	35.6

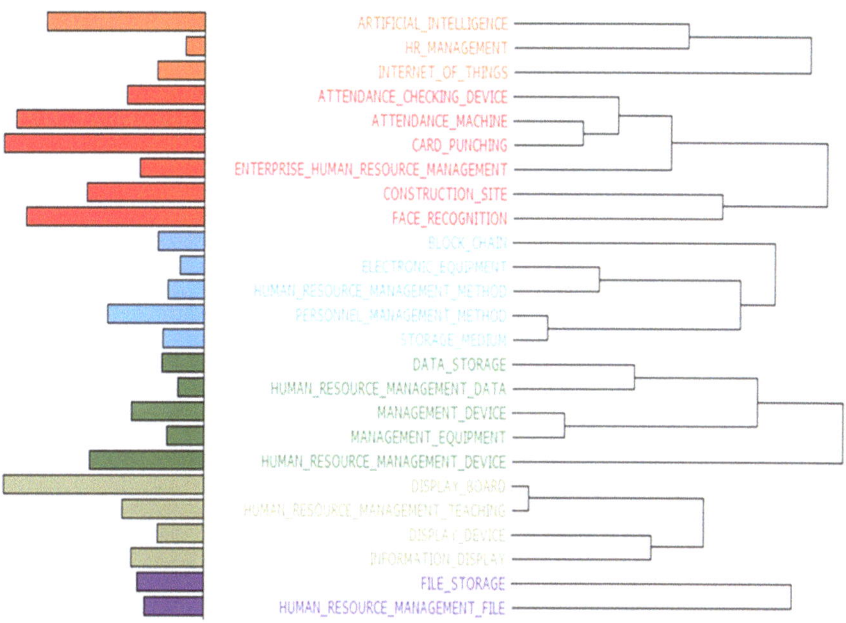

Fig. 7 Six clusters of HRM patent family phrases (*Source* Authors, 2023)

Figures 8, 9, 10, 11, 12 and 13 present six network graphs indicating which phrases co-occurred most often within each of the clusters:

- Cluster 1: The phrases that occur the most often with the phrase artificial intelligence are mostly HR management, storage medium and Internet of Things;

- Cluster 2: The phrases that occur the most often with the phrase card punching are mostly related to the devices such as attendance machines and attendance checking devices, and methods such as face recognition;
- Cluster 3: The phrase personnel management method is related to devices such as storage medium, and electronic purposes as well as some specific application areas such as construction site, face recognition, artificial intelligence and Internet of Things;
- Cluster 4: Phrase human resource management device often occurs with the phrase management device, blockchain, card punching, electronic equipment and attendance machine;
- Cluster 5: The phrases that occur often with the phrase display board are related to teaching such as human resource management teaching and displaying such as display devices, information display and management equipment; and
- Cluster 6: Phrase file storage co-occurs most often with the phrase human resources management file.

Overall, this study was conducted to better understand the level of innovation in the area of HRM and the methodology that was employed to examine HRM-related innovation provided significant insights into the patent landscape, geographical distribution, and technical substance of HRM-related patents.

The first stage of data collecting included a rigorous search procedure, including an exhaustive compilation of synonyms and associated keywords pertaining to HRM. The use of this methodology is crucial, since the exploration of innovation within

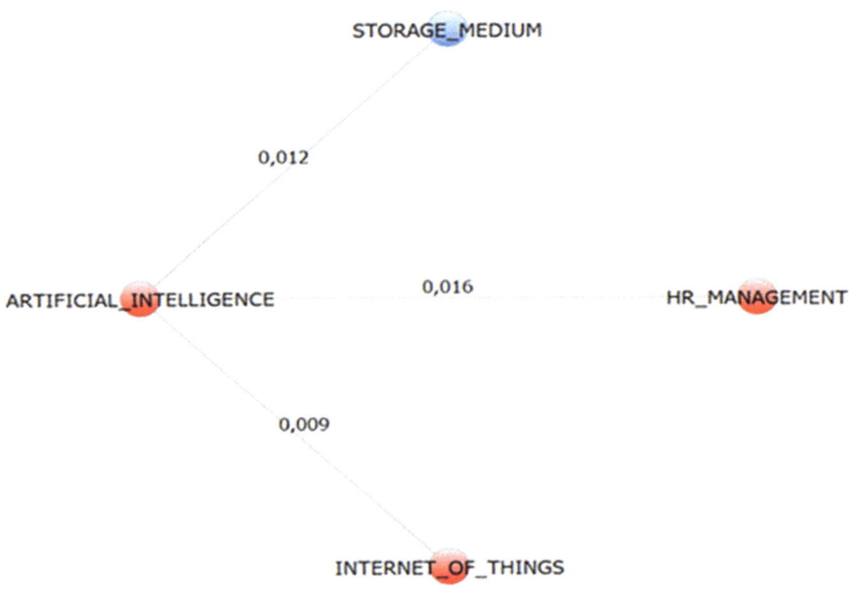

Fig. 8 Circular graph for cluster 1 (*Sources* Authors' work, 2023)

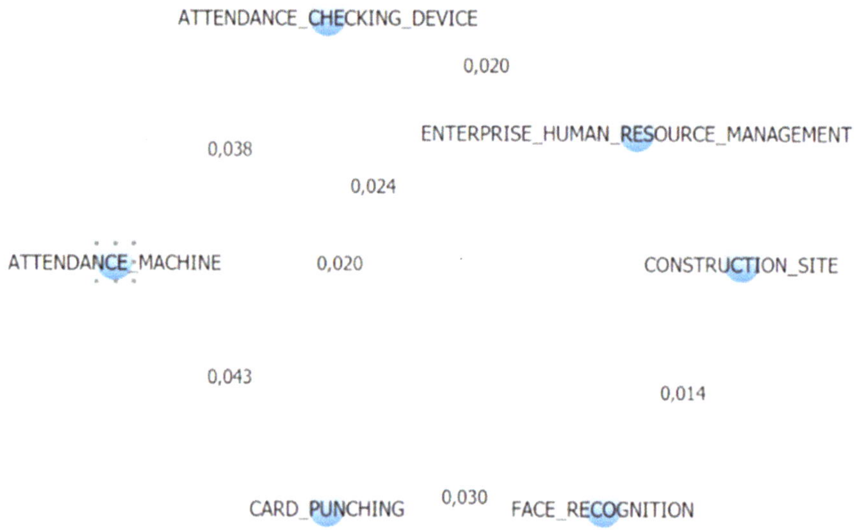

Fig. 9 Circular graph for cluster 2 (*Sources* Authors' work, 2023)

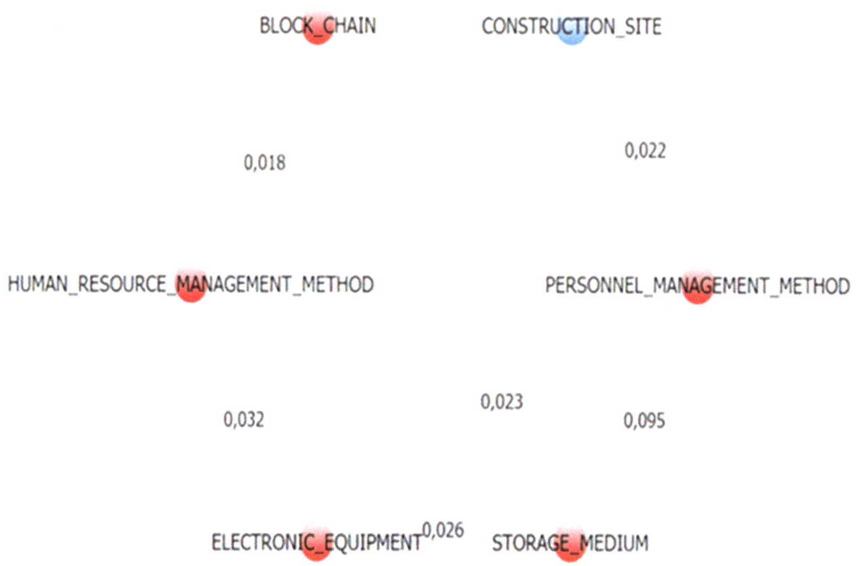

Fig. 10 Circular graph for cluster3 (*Sources* Authors' work, 2023)

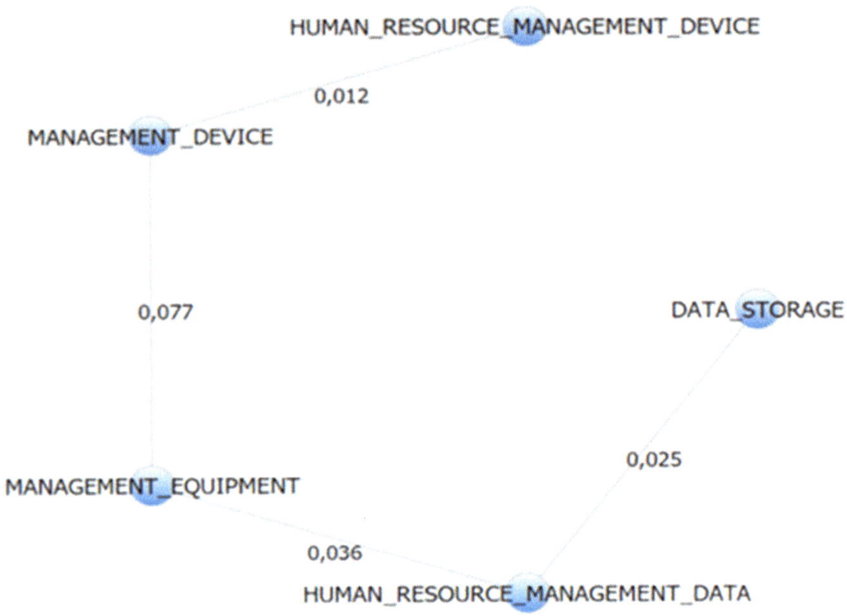

Fig. 11 Circular graph for cluster 4 (*Sources* Authors' work, 2023)

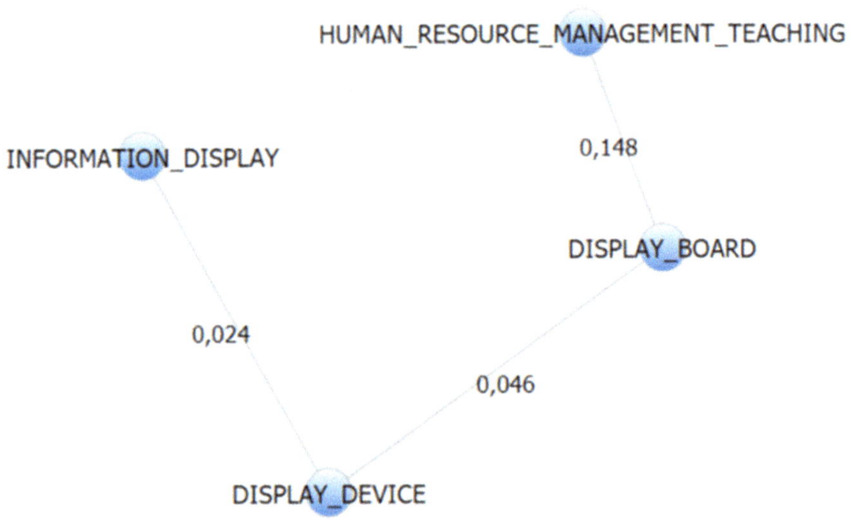

Fig. 12 Circular graph for cluster 5 (*Sources* Authors' work, 2023)

Fig. 13 Circular graph for cluster 6 (*Sources* Authors' work, 2023)

HRM may be articulated using several terminologies and expressions, so guaranteeing that potentially significant patents are not disregarded. The use of spelling variants, such as "labor" and "labour", serves to increase the comprehensiveness of the search. In addition, the analysis included the differentiation between patents classified as "active" and those classified as "inactive" based on their legal standing, therefore offering a thorough assessment of the present patent environment while recognizing its historical underpinnings.

As promised in the introduction, this chapter provided answers to the stated four research questions: (RQ1) What is the time-wise publication trend in innovation activities related to HRM?; (RQ2) Who are the top assignees and inventors within the technical field related to human resources management?; (RQ3) How are innovation activities related to the HRM spread across different countries?; and (RQ4) What is the technical content of HRM innovations?

First, as the answer to the RQ1, the results show a significant increase in the total number of HRM patent families starting in 2017 and peaking in 2021 with 481 HRM patent families. However, there was a slight decrease in patenting activities in 2022. The data suggests that there has been a recent surge in HRM innovation and technological advancement, especially in 2021 and 2022. This trend may be driven by the need for more efficient human resources processes, possibly in response to challenges posed by the COVID-19 pandemic. The publication timeline for HRM patents across the top five countries also indicates that the highest number of HRM patents for these countries were published in recent years, particularly in 2021 and 2022.

Second, as the answer to RQ2, the results identify China as the leader in patenting activities related to HRM, with almost 77% of the total number of HRM patent families. Key players in HRM innovation are primarily from China and Japan, and they come from both business and academia. This suggests that China is a dominant force in HRM innovation, with active participation from various entities in both countries.

Third, as the answer to RQ3, the geographical distribution of HRM patents shows that China's assignees lead in patenting activities, followed by India, Japan, the United States of America, and South Korea. However, Spain, Romania, Ukraine, and Singapore each have only a minimal presence in HRM patenting. This distribution highlights China's prominent role in HRM innovation, with other countries contributing to a lesser extent.

Finally, as the answer to RQ4, it has been shown that the technical content of HRM innovations is diverse and spans multiple IPC sections and subdomains. HRM patent families were registered under eight IPC sections, indicating that HRM is an interdisciplinary field. These sections include physics, human necessities, performing operations, transport, electricity, mechanical engineering, fixed constructions, chemistry, metallurgy, and textiles. This diversity reflects the multifaceted nature of HRM, which includes tasks related to both management and organization (e.g., physics and human needs) and the development and implementation of technological solutions (e.g., electrical and mechanical engineering). Additionally, the technical content analysis of HRM patents revealed key categories with significant patent family numbers, such as data processing systems, electrical digital data processing, and recognition of data. These categories suggest that HRM has become increasingly automated and data-driven, with technologies related to data processing and management playing a substantial role. There is also a focus on efficiency, automation, document management, and communication within HRM, as indicated by patent categories related to office automation, storage solutions, and educational appliances.

Additionally, text mining of patent abstracts unveiled key phrases frequently occurring in HRM patents. These phrases, such as "management device", "personnel management method", "artificial intelligence", and "file storage", shed light on the core themes of HRM innovation. The importance of these phrases was quantified using the TF-IDF metric, emphasizing their significance in the HRM patent landscape. Cluster analysis further revealed associations between these phrases, offering insights into the relationships and patterns within HRM innovation. For instance, the cluster containing "artificial intelligence" indicated a strong connection with HRM and the Internet of Things, highlighting the growing role of artificial intelligence in HRM.

5 Conclusion

This study conducted an in-depth analysis of HRM patents, yielding several significant findings. After data preparation, 2109 records related to HRM were inspected, with the majority (57%) found to be active patents, and the rest inactive due to reasons like withdrawal, rejection, nonpayment, or expiration. China emerged as the dominant player in HRM patenting activities, contributing to nearly 77% of the total HRM patent families, followed by India, Japan, the United States, and South Korea.

The research also revealed that the highest number of HRM patents across these top five countries was published in the years 2021 and 2022, indicating a recent surge in patenting activities. Notably, key players in the HRM technological field predominantly hailed from China and Japan, encompassing both business and academic entities. The technological trends unveiled various subdomains with significant patent activity, including IT methods for management, furniture, games, control, audio-visual technology, and other consumer goods, each with distinct growth patterns over the years. Furthermore, a detailed examination of China's patenting activities showcased shifts in focus, with a pronounced increase in patents related to furniture and games in 2020 and 2021. Lastly, the analysis delved into the content of HRM patents, identifying key categories such as data processing systems, time or attendance registers, office furniture, educational and demonstration appliances, and display systems. Text mining of patent abstracts unearthed frequently occurring phrases like "management device", "personnel management method", "artificial intelligence", and "file storage". Cluster analysis unveiled six clusters of co-occurring phrases, offering insights into thematic relationships within HRM patents.

However, although this research offers a thorough examination of the patent landscape in the HRM field and provides valuable insights into the geographical distribution, technical patterns, major contributors, and patent content, thereby emphasizing the ever-evolving nature of innovation in HRM, the authors are aware of its limitations. Firstly, it primarily focused on patent data, which may not capture all forms of innovation in HRM. Qualitative research methods, surveys, and interviews could complement the patent-based analysis to provide a more holistic view of HRM innovation. Additionally, the study did not delve into the quality or impact of the patents, which could be a valuable area for future investigation.

Future research in the domain of HRM innovation could take several directions. Firstly, an exploration of the practical applications and impact of the identified innovation themes could provide valuable insights into the real-world significance of these patents. Secondly, examining the collaborative networks and knowledge flows among key players in HRM innovation could offer a deeper understanding of the innovation ecosystem. Lastly, given the dynamic nature of technology, continuous monitoring of patent trends and regular updates to this analysis would provide an up-to-date perspective on HRM innovation.

Acknowledgements The authors of this chapter would like to thank professor Mirjana Pejić Bach for introducing them to the topic of patent analysis.

References

Abood, A., & Feltenberger, D. (2018). Automated patent landscaping. *Artificial Intelligence and Law, 26*(2), 103–125.

Ackah, A. P. (2017). The Core critical areas for innovation to sustain business growth: Management Implications and Business Outputs. *The International Journal of Management, 3*(2), 183–195.

Adel, M. E. (2021). Zipf's law applications in patent landscape analysis. *World Patent Information, 64*, 102012.

Aggarwal, S., & Chandra, A. (2021). An insight into patent landscape analysis of plant stem cells. *World Patent Information, 65*, 102025.

Aithal, P. S., & Aithal, S. (2018). Patent analysis as a new scholarly research method. *International Journal of Case Studies in Business, IT and Education, 2*(2), 33–47. https://doi.org/10.5281/Zenodo.1404184

Alam, S., Munizu, M., & Jillbert, J. (2019). Effect of use of information technology on innovation capability, competitiveness, and firm performance: Case of manufacturing industry in South Sulawesi. In *IOP conference series: Earth and environmental science* (Vol. 235, No. 1, p. 012008). IOP Publishing.

Allison, J. R., & Lemley, M. A. (2000). Who's patenting what-an empirical exploration of patent prosecution. *Vanderbilt Law Review, 53*, 2099.

Amarakoon, U., Weerawardena, J., & Verreynne, M. L. (2018). Learning capabilities, human resource management innovation and competitive advantage. *The International Journal of Human Resource Management, 29*(10), 1736–1766.

Androshchuk, H. O., & Kvasha, T. K. (2019). Patent landscape as a tool for forecasting world technological trends: Arms and military equipment. *Science, Technologies, Innovations, 4*(12), 28–40.

Anwar, G., & Abdullah, N. N. (2021). The impact of human resource management practice on organizational performance. *International Journal of Engineering, Business and Management (IJEBM), 5*.

Armstrong, M. (2006). *A handbook of human resource management practice*. London, UK: Kogan Page Publishers.

Azis, F. A., Rijal, M., Suhaimi, H., & Abas, P. E. (2022). Patent landscape of composting technology: A review. *Inventions, 7*(2), 38.

Azizi, M. R., Atlasi, R., Ziapour, A., Abbas, J., & Naemi, R. (2021). Innovative human resource management strategies during the COVID-19 pandemic: A systematic narrative review approach. *Heliyon, 7*(6).

Baars, H., & Kemper, H. G. (2008). Management support with structured and unstructured data–an integrated business intelligence framework. *Information Systems Management, 25*(2), 132–148. https://doi.org/10.1080/10580530801941058

Bamakan, S. M. H., Bondarti, A. B., Bondarti, P. B., & Qu, Q. (2021). Blockchain technology forecasting by patent analytics and text mining. *Blockchain: Research and Applications, 2*(2), 100019.

Bircan, I., & Gencler, F. (2015). Analysis of innovation-based human resources for sustainable development. *Procedia-Social and Behavioral Sciences, 195*, 1348–1354.

Birkinshaw, J., Hamel, G., & Mol, M. (2008). Management innovation. *Academy of Management Review, 33*, 825–845.

Bosilj Vukšić, V. (2020). Informacijska tehnologija i poslovna etika. In M. Pejić Bach & M. Spremić (ur.), *Osnove poslovne informatike*. Ekonomski fakultet Zagreb.

Business Research Company. (2023). *Information technology global market report 2023*. Retrieved August 30, 2023, from https://www.thebusinessresearchcompany.com/report/information-technology-global-market-report

Choi, D., & Song, B. (2018). Exploring technological trends in logistics: Topic modeling-based patent analysis. *Sustainability, 10*(8), 2810.

Chun, E., Jun, S., & Lee, C. (2021). Identification of promising smart farm technologies and development of technology roadmap using patent map analysis. *Sustainability, 13*(19), 10709.

Chung, J., Ko, N., Kim, H., & Yoon, J. (2021). Inventor profile mining approach for prospective human resource scouting. *Journal of Informetrics, 15*(1), 101103.

Cruz, A. C., Al Khalil, A., Kirbac, A., Cowan, K., & Ayyildiz, O. (2022, August). Forecasting the use of decontamination robots in the COVID era. In *2022 Portland International Conference on Management of Engineering and Technology (PICMET)* (pp. 1–9). IEEE.

Daim, T. U., Rueda, G., Martin, H., & Gerdsri, P. (2006). Forecasting emerging technologies: Use of bibliometrics and patent analysis. *Technological Forecasting and Social Change, 73*(8), 981–1012.

De Leede, J., & Looise, J. K. (2005). Innovation and HRM: Towards an integrated framework. *Creativity and Innovation Management, 14*(2), 108–117.

Dede, N. P. (2019). Enhancing employee innovative work behavior through human resource management practices. In *Handbook of research on managerial thinking in global business economics* (pp. 1–21). IGI Global.

Dougherty, D. (1992). A practice-centered model of organizational renewal through product innovation. *Strategic Management Journal, 13*(S1), 77–92.

Dutta, D. (2018). Social media and technology trends in HRM: Cases in recruitment and talent management. In *Social media and journalism-trends, connections, implications*. IntechOpen.

Ejermo, O., & Karlsson, C. (2006). Interregional inventor networks as studied by patent coinventorships. *Research Policy, 35*(3), 412–430.

Elliott, G. (2007). Basic of US patents and the patent system. *The AAPS Journal, 9*, E317–E324.

Elvers, D., Song, C. H., Steinbüchel, A., & Leker, J. (2016). Technology trends in biodegradable polymers: Evidence from patent analysis. *Polymer Reviews, 56*(4), 584–606.

Emma, P. G. (2005). What is patentable? *IEEE Micro, 25*(4), 7–9. https://doi.org/10.1109/MM.2005.78

European Commission. (1995). *Green paper on innovation*. http://europa.eu/documents/comm/green_papers/pdf/com95_688_en.pdf

Fareri, S., Apreda, R., Mulas, V., & Alonso, R. (2023). *The energy worker profiler from technologies to skills to realize energy efficiency in manufacturing*. arXiv:2301.09445

Fatimi, A. (2022). Chitosan-based hydrogels: Patent analysis. *Materials Proceedings, 9*(1), 1.

Forestal, R. L. (2023). Using automated patent landscaping and legal geography analysis to spot biopiracy activities in the island of Hispaniola. *World Patent Information, 72*, 102174.

Freeman, C., & Perez, C. (1988). Structural crises of adjustment: business cycles and investment behaviour. In G. Dosi, C. Freeman, R. Nelson, G. Silverberg, & L. Soete (Eds.), *Technical Change and Economic Theory* (pp. 38–66). London, UK: Pinter Publishers.

Garcia, R., & Calantone, R. (2002). A critical look at technological innovation typology and innovativeness terminology: A literature review. *Journal of Product Innovation Management, 19*, 110–132.

Hall, B. H. (2007). Patents and patent policy. *Oxford Review of Economic Policy, 23*(4), 568–587.

Hamilton, W., & Till, I. (1948). What is a patent. *Law and Contemporary Problems, 13*, 245.

Harhoff, D., & Wagner, S. (2009). The duration of patent examination at the European Patent Office. *Management Science, 55*(12), 1969–1984.

Henderson, R. M., & Clark, K. B. (1990). Architectural innovation: The reconfiguration of existing product technologies and the failure of established firms. *Administrative Science Quarterly, 35*(1), 9–30.

Hsu, D. H., & Ziedonis, R. H. (2013). Resources as dual sources of advantage: Implications for valuing entrepreneurial-firm patents. *Strategic Management Journal, 34*(7), 761–781.

Jolly, A. (2008). *The growing business handbook* (10th ed.). Kogan Page.

Kiesler, S., Siegel, J., & McGuire, T. W. (1984). Social psychological aspects of computer-mediated communication. *American Psychologist, 39*, 1124–1134.

Kim, G., & Bae, J. (2017). A novel approach to forecast promising technology through patent analysis. *Technological Forecasting and Social Change, 117*, 228–237.

Kim, K. H., Han, Y. J., Lee, S., Cho, S. W., & Lee, C. (2019). Text mining for patent analysis to forecast emerging technologies in wireless power transfer. *Sustainability, 11*(22), 6240.

King, J. L., Gurbaxani, V., Kraemer, K. L., McFarlan, F. W., Raman, K. S., & Yap, C. S. (1994). Institutional factors in information technology innovation. *Information Systems Research, 5*(2), 139–169.

Klongthong, W., Muangsin, V., Gowanit, C., & Muangsin, N. (2021). A patent analysis to identify emergent topics and convergence fields: A case study of chitosan. *Sustainability, 13*(16), 9077.

Kultti, K., Takalo, T., & Toikka, J. (2007). *Patents hinder collusion.* University of Helsinki.

Kwon, K., Jun, S., Lee, Y. J., Choi, S., & Lee, C. (2022). Logistics technology forecasting framework using patent analysis for technology roadmap. *Sustainability, 14*(9), 5430.

Le Bas, C., Mothe, C., & Nguyen Thi, T. U. (2011). Technological innovation persistence: Literature survey and exploration of the role of organizational innovation. *Available at SSRN 1969293.* https://doi.org/10.2139/ssrn.1969293

Lee, S., Lee, S., Seol, H., & Park, Y. (2008). Using patent information for designing new product and technology: Keyword based technology roadmapping. *R&D Management, 38*(2), 169–188.

Lee, S., Yoon, B., Lee, C., & Park, J. (2009). Business planning based on technological capabilities: Patent analysis for technology-driven roadmapping. *Technological Forecasting and Social Change, 76*(6), 769–786.

Li, Z. Y., & Meyer, W. (2014). Patents: A tool to bring innovation from the lab bench to the marketplace. *Yeast Metabolic Engineering: Methods and Protocols*, 297–310.

Loewenberger, P. (2016). Human resource development, creativity and innovation. *Human resource management, innovation and performance* (pp. 48–65). Palgrave Macmillan.

Ma, M. Y. (2015). *Fundamentals of patenting and licensing for scientists and engineers.* World Scientific.

Machado, C., & Davim, J. P. (2014). *Human resource management and technological challenges.* Springer International Publishing.

Maghsoudi, M., Nourbakhsh, R., Kermani, M. A. M., & Khanizad, R. (2023). *The power of patents: Leveraging text mining and social network analysis to forecast IoT trends.* arXiv:2309.00707

Mao, S., Hu, Y., Yuan, X., Zhang, M., Qiu, Q., & Wu, P. (2022). Analysis of patent application attention: A network analysis method. *Frontiers in Physics, 10*, 893348.

Mathis, R. L., Jackson, J. H., & Valentine, S. R. (2015). *Human resource management: Essential perspectives.* Cengage Learning.

Moehrle, M. G., Walter, L., Geritz, A., & Müller, S. (2005). Patent-based inventor profiles as a basis for human resource decisions in research and development. *R&D Management, 35*(5), 513–524.

Mushkudiani, Z., Gechbaia, B., Gigauri, I., & Gulua, E. (2020). Global, economic and technological trends in human resource management development. *Access Journal, 1*(1), 53–60.

Naidoo, I. P., & Hoque, M. (2018). Impact of information technology on innovation in determining firm performance. *African Journal of Science, Technology, Innovation and Development, 10*(6), 643–653.

National Research Council. (1988). *The national challenge in computer science and technology: A Report of the National Research Council.* Washington, D.C., USA: National Academy Press.

National Research Council. (1994). *Information technology in the service society: A Twenty-First Century Leverage.* Washington, D.C., USA: National Academy Press.

Nomaler, Ö., & Verspagen, B. (2021). *Patent landscaping using "green" technological trajectories.* Maastricht Economic and Social Research Institute on Innovation and Technology (UNU-MERIT).

Opatha, H. H. D. N. P. (2009). *Human resource management.* University of Sri Jayewardenepura.

Park, S., & Jun, S. (2022). Patent analysis using Bayesian data analysis and network modeling. *Applied Sciences, 12*(3), 1423.

PatSeer. (2023). http://patseer.com/

Pawar, Y. S., & Shaikh, S. A. (2021). Patent literature as a tool for talent acquisition–A case study. *Allana Management Journal of Research, Pune, 11*, 8–18.

Phadnis, R., & Hirwani, R. R. (2005). Patent analysis as a tool for research planning: Case study of phytochemicals in tea. *Journal of Intellectual Property Rights, 10*(3), 221–231.

Porter, M. E. (1990). The competitive advantage of nations. *Harvard Business Review, March–April 1990*, 73–93.

Provalis Research. (2021). *WORDSTAT 9 text analytics software user's guide.* https://provalisresearch.com/Documents/WordStat9.pdf

Przytuła, S., Strzelec, G., & Krysińska-Kościańska, K. (2020). Re-vision of future trends in human resource management (HRM) after COVID-19. *Journal of Intercultural Management, 12*(4), 70–90.

Radu, L. D. (2015). Investments in technological innovations: A literature review of organization determinants. *European Scientific Journal, ESJ, 11*(3), 52–59.

Rogers, E. M. (1995). Diffusion of innovations: Modifications of a model for telecommunications. *Die diffusion von innovationen in der telekommunikation*, 25–38.

Rosenberg, P. D. (1992). *Patent law basics*. Clark, Boardman, Collagham.

Sadiku, M. N. (1997). Patents in general. *IEEE Potentials, 16*(4), 23–24.

Sajjad, M., Riaz, A., Chani, M., & Hussain, R. (2020). Innovations in human resources management: Mediating role of intrinsic motivation. https://doi.org/10.21272/MMI.2020.1-08

Sichelman, T. (2009). Commercializing patents. *Stanford Law Review, 62*, 341–412.

Silge, J., & Robinson, D. (2017). *Text mining with R: A tidy approach*. O'Reilly Media.

Singh, A., & Kumar, S. (2022). Identifying innovations in human resources: Academia and industry perspectives. In *Research anthology on human resource practices for the modern workforce* (pp. 219–231). IGI Global.

Singh, V., Chakraborty, K., & Vincent, L. (2016). *Patent database: Their importance in prior art documentation and patent search*.

Sinha, M., & Pandurangi, A. (2016). *Guide to practical patent searching and how to use PatSeer for patent search and analysis* (2nd ed.). Gridlogic Technologies Pvt. Ltd.

Soni, L., Tripathi, S. K., & Musyuni, P. (2013). Overview of patent landscaping a tool for research. *Pharma Science Monitor, 4*(1), 3565–3583.

Steijn, M. P. A. (2021). Improvement on the association strength: Implementing a probabilistic measure based on combinations without repetition. *Quantitative Science Studies, 2*(2), 778–794.

Stewart, G. L., & Brown, K. G. (2019). *Human resource management*. Wiley.

Stone, D. L., & Deadrick, D. L. (2015). Challenges and opportunities affecting the future of human resource management. *Human Resource Management Review, 25*(2), 139–145.

Stone, R. J., Cox, A., & Gavin, M. (2020). *Human resource management*. Wiley.

Sun, J. Y. (2013). Innovations of human resource management in new knowledge economy times. In *The 19th International Conference on Industrial Engineering and Engineering Management: Management System Innovation* (pp. 753–759). Springer.

Tang, J., Wang, B., Yang, Y., Hu, P., Zhao, Y., Yan, X., Gao, B., Huang, M., Xu, P., Li, W., & Usadi, A. K. (2012). Patentminer: Topic-driven patent analysis and mining. In *Proceedings of the 18th ACM SIGKDD International Conference on Knowledge Discovery and Data Mining* (pp. 1366–1374).

Trippe, A. (2015). Guidelines for preparing patent landscape reports. *Patent landscape reports*. WIPO.

Tseng, Y. H., Lin, C. J., & Lin, Y. I. (2007). Text mining techniques for patent analysis. *Information Processing & Management, 43*(5), 1216–1247.

Tyczewska, A. (2014). Patents—An important added value for economy and science. *BioTechnologia. Journal of Biotechnology Computational Biology and Bionanotechnology, 95*(4).

van Rijn, T., & Timmis, J. K. (2023). Patent landscape analysis—Contributing to the identification of technology trends and informing research and innovation funding policy. *Microbial Biotechnology, 16*(4), 683–696.

Vlad, F. (2010). *Patent landscaping—As a portfolio strength indicator*.

Vrontis, D., Christofi, M., Pereira, V., Tarba, S., Makrides, A., & Trichina, E. (2022). Artificial intelligence, robotics, advanced technologies and human resource management: A systematic review. *The International Journal of Human Resource Management, 33*(6), 1237–1266.

Walker, R. D. (1995). *Patents as scientific and technical literature*. Scarecrow Press.

White, M. J. (2009). Patents and patent searching. In *Encyclopedia of library and information sciences* (pp. 4114–4126). CRC Press.

Wich, Y., Warschat, J., Spath, D., Ardilio, A., König-Urban, K., & Uhlmann, E. (2013, July). Using a text mining tool for patent analyses: Development of a new method for the repairing of gas turbines. In *2013 Proceedings of PICMET'13: Technology Management in the IT-Driven Services (PICMET)* (pp. 1010–1016). IEEE.

Willinger, M., & Zuscovitch, E. (1988). Towards the economics of information-intensive production systems: the case of advanced materials. In G. Dosi, C. Freeman, R. Nelson, G. Silverberg, & L. Soete (Eds.), *Technical change and economic theory* (pp. 239–255). London, UK: Pinter Publishers.

Wilson, R. M. (1987). Patent analysis using online databases—I. Technological trend analysis. *World Patent Information, 9*(1), 18–26.

wipo.int. (2023). *Innovation and intellectual property.* https://www.wipo.int/ip-outreach/en/ipday/2017/innovation_and_intellectual_property.html

Wolfe, R. A. (1995). Human resource management innovation: Determinants of their adoption and implementation. *Human Resource Management, 34*, 313–327.

Wolfe, R., Wright, P., & Smart, D. (2006). Radical HRM innovation and competitive advantage: The Money ball story. *Human Resource Management, 45*, 111–145.

Wood, S. (1999). Human resource management and performance. *International Journal of Management Reviews, 1*(4), 367–413.

World Intellectual Property Organization (WIPO). (2023). *Guide to the international patent classification* (2023 ed.). https://www.wipo.int/edocs/pubdocs/en/wipo-guide-ipc-2023-en-guide-to-the-international-patent-classification-2023.pdf

Yang, Y. Y., Akers, L., Yang, C. B., Klose, T., & Pavlek, S. (2010). Enhancing patent landscape analysis with visualization output. *World Patent Information, 32*(3), 203–220.

Zeng, Z., & Qi, L. (2021). "Internet+ artificial intelligence" human resource information management system construction innovation and research. *Mathematical Problems in Engineering, 2021*, 1–11.

Zhou, W., & Wang, X. (2021). Human gene therapy: A patent analysis. *Gene, 803*, 145889.

Zhu, L., Zhu, D., Wang, X., Cunningham, S. W., & Wang, Z. (2019). An integrated solution for detecting rising technology stars in co-inventor networks. *Scientometrics, 121*, 137–172.

Artificial Intelligence (AI) in Human Resource Management (HRM)

Hafinas Halid, Kamalesh Ravesangar, Syaza Lyana Mahadzir, and Siti Norjannah Abd Halim

Abstract Organizations are increasingly confronted with the need to modernize and re-invent themselves in order to keep abreast of the ever-evolving trends in social media and information technology. As a result, the adoption of AI technologies is on the rise. The integration of Artificial Intelligence into businesses is bringing computers and people closer together, as well as exploring ways to use AI to increase productivity, ease of use, and efficiency. HR professionals are increasingly focusing on the importance of optimizing the combination of manual and automated work for a user-friendly work environment. As companies continue to incorporate AI into their HR processes at varying rates, it is essential for HR professionals to be aware of and prepared for these technological advancements. Human resource (HR) technology that is based on Artificial Intelligence can be beneficial to a business in the short and long run. This book chapter aims to provide an overview of the use of Artificial Intelligence in Human Resource Management (HRM) practices in the digital era. Additionally, it aims to provide an understanding of the Use of AI-based Software in HRM, as well as the opportunities and challenges in the digital world related to AI in HRM.

Keywords Artificial Intelligence · Technology · HRM practices · Digital HRM · Machine learning

---

H. Halid (✉)
School of Business Management, Universiti Utara Malaysia, Kedah, Malaysia
e-mail: hafinas@uum.edu.my

K. Ravesangar
Tunku Abdul Rahman University of Management and Technology (TARUMT), Kuala Lumpur, Penang, Malaysia

S. L. Mahadzir · S. N. A. Halim
Albukhary International University, Alor Setar, Kedah, Malaysia

C. Machado (ed.), *Building the Future with Human Resource Management*, Management and Industrial Engineering, https://doi.org/10.1007/978-3-031-52811-8_2

1 Introduction

The emergence of Artificial Intelligence (AI) has significantly altered how computer systems interact with tasks that mimic human thinking. This transformation has created fresh opportunities for companies to improve operational efficiency. AI, as defined by Kaplan and Haenlein (2019), is the term used to describe computerized systems that are capable of carrying out both cognitive and physical tasks, solving challenging puzzles, and making decisions without direct human guidance. By providing creative ways to harness the power of AI, this idea has transformed numerous industries and paved the way for greater productivity and optimized operations.

In the landscape of modern research, AI has taken on an unprecedented role. Tang et al. (2022) emphasize that AI has become a focal point of exploration across academic and industrial sectors, permeating various fields of study. Due to its transformational potential, it has gained importance across a wide range of fields, expanding its relevance beyond only technology. The adoption of AI is no longer restricted to particular industries; rather, it has evolved into an interdisciplinary force that influences how businesses run, issues are resolved, and decisions are taken. The application of AI across a variety of industries highlights how important it is as a catalyst for creativity, efficiency, and advancement.

1.1 The Evolution of Artificial Intelligence (AI)

The 1950s marked the beginning of artificial intelligence as a subject of study (Jatobá et al., 2019). The past few decades have seen the rapid expansion and revolutionary advancement of Artificial Intelligence (AI), which has profoundly changed how we interact with technology and changed numerous worldwide sectors. AI applications have now smoothly incorporated into our daily lives after beginning as abstract ideas in the middle of the twentieth century, demonstrating amazing improvements in AI research, algorithms, and processing power. With the potential for huge advancements in fields like deep learning, natural language processing, and autonomous systems, the future of AI offers even more promise for revolutionizing industries like healthcare, education, and business.

The current environment, which is characterized by the proliferation of AI-driven technologies that have already touched numerous sectors, is evolving as AI. At the heart of this change are machine learning algorithms, neural networks, and data-driven insights, enabling previously unheard-of capabilities in processes like decision-making, language translation, and picture recognition. Although AI-powered technologies have greatly improved productivity and convenience, they also create serious moral and societal issues. A closer examination of AI's possible effects on the labor market and human society, as well as increasing regulation and monitoring, are expected to be part of the future trajectory of the technology.

1.2 Artificial Intelligence (AI) in Business

Artificial Intelligence (AI) is revolutionizing the way businesses operate, enabling them to make more informed decisions, increase productivity, and reduce expenses. Machine learning algorithms have been a popular application of AI in recent years, allowing businesses to analyze large amounts of data and gain insights that were previously inaccessible. In particular, AI has the potential to simplify processes, reduce mistakes, and enhance customer experiences. AI applications encompass the recognition of machines and the utilization of language. Concepts are formed into abstractions and used to solve intricate issues. The ever-increasing amount of data available and the steady progression of computational power and systems have led to the development of various AI applications across a wide range of industries. Although the classifications and concepts of AI vary depending on the objectives and scope, the primary characteristics of AI are the replication of human reasoning purposes, primarily knowledge, and the resolution of problems. The use of AI has been widely asserted as a disruptive force for businesses around the world and across a broad spectrum of industries. Organizations that implement AI applications are projected to experience benefits in terms of increased business value, cost savings, and enhanced business productivity (Mikalef & Gupta, 2021). As a result, many organizations invest in AI technologies in the search for a competitive edge.

1.2.1 Applications of Artificial Intelligence (AI) in Business

Artificial Intelligence (AI) in business is a collection of computational technologies and processes that enable organizations to extract meaningful insights from data, leading to improved operational efficiency, decreased expenses, increased revenue sources, and the attainment of corporate, societal, and economic goals. Here are some artificial intelligence applications in business.

(a) **Invoices and Bills**

Every business has a financial obligation that must be managed. Accounting and bill-making software can be a time-consuming and laborious process, and errors in the math can lead to significant losses. AI has made it easier and more accurate to manage money by automating tasks. There are a variety of accounting software types available on the market, which offer features such as invoice data extraction, segregation, and scanning, uploading, and storing of paper invoices. Electronic invoices are simpler to manage due to their automated reading and storage. AI-powered accounting tools are highly accurate and organized, making managing money a breeze.

(b) **Social Media**

The utilization of social media platforms has become a highly effective marketing tool for businesses, allowing them to showcase their offerings to a broad audience. If a company is well-versed in the use of social media platforms, it is likely to attract a large number of potential customers. However, due to the large number of users, it can be difficult for a company to collect customer feedback directly. AI can be utilized to gain insight into a brand's position in the market, as well as to gain a deeper understanding of its customers. This information can then be used to develop a plan, enhance its social media presence, and ultimately build the brand.

(c) **Customer Analysis**

Customers are essential to any business, and their decisions can have a major impact on the success of any product. To maximize customer engagement and improve performance in any field, businesses must first conduct a comprehensive analysis of their customer base. In the past, this was difficult due to the fact that most customer interactions took place in person and feedback was based on manual forecasting or emotional factors. However, with the advent of AI, businesses can now conduct comprehensive consumer surveys that can reveal areas of improvement beyond superficial trends. AI provides reliable information and assists in the implementation of initiatives to increase engagement and facilitate sales through improved customer service. This is why AI is beneficial for businesses: it enables them to become more customer-focused.

(d) **Cyber Security**

The internet has revolutionized the storage and administration of any business, however, this has also increased the risk of data theft and exposure. Every firm must take online security measures, as the majority of its essential databases, including financial data, business strategies, and personal information, are stored online. AI is now being used in the cybersecurity sector, which is essential for any business. Cyber professionals are now able to understand and remove any unwanted noise or data with the help of AI. This is beneficial for businesses, as it allows them to stay informed of any unusual activity or viruses, and to be prepared for any potential attack. Additionally, AI analyses large amounts of data and adapts the system accordingly to reduce the likelihood of cyberattacks.

1.2.2 The Significance of Artificial Intelligence (AI) in Business

The utilization of technology has had a profound impact on human life. It has provided a wide range of precise solutions and other benefits to humanity. Artificial Intelligence (AI) has brought a range of features to the industry, such as predictive action, rapid processing, and enhanced insight. AI is increasingly being adopted in all industries, both in order to develop its core features and to enhance the existing features in order to improve the utilization of technology for the benefit of both industries and society. As technology advances, the importance of AI in business has grown. From automating mundane tasks to delivering personalized experiences, there is no doubt that AI is revolutionizing the business landscape. In this article, we will look at how AI is transforming the way we conduct business and the advantages it offers to businesses of all sizes.

(a) **Automation of Repetitive Tasks**

One of the primary benefits of AI is its capacity to automate mundane, low-priority tasks, allowing employees to dedicate more time to more complex and strategic initiatives. From data entry to customer service to financial analysis, AI is capable of completing these tasks quickly and accurately, enabling your team to concentrate on higher-priority tasks that necessitate human ingenuity and problem-solving abilities. For instance, with the integration of AI-driven chatbots, many of these interactions could be automated, allowing the company to dedicate more time to more complex matters that necessitate a human touch.

(b) **Personalization**

The importance of personalization in customer experience cannot be overstated, and AI is enabling businesses to personalize their customer experience in novel and innovative ways. Through the analysis of customer data and behavioral patterns, AI can personalize customer experiences, providing personalized recommendations and advertisements tailored to individual preferences. From product recommendations on e-commerce sites to targeted advertising on social media platforms, AI is enabling businesses to create truly personalized customer experiences.

(c) **Predictive Maintenance**

AI is revolutionizing the way businesses operate by enabling them to anticipate the occurrence of malfunctions in machinery or equipment. By analyzing sensors and other data, AI can forecast when equipment will require maintenance, allowing companies to plan for the future and minimize the need for costly downtime (Vrignat et al., 2022). Ultimately, AI is transforming the way we conduct business, offering numerous advantages and possibilities for companies to optimize processes, enhance productivity, and enhance customer experiences. Integrating AI into businesses can help them remain competitive in the rapidly changing market and ensure a prosperous future.

(d) **Fraud Detection**

Fraud is a serious issue that affects businesses of all types, but AI is helping to reduce the likelihood of financial losses and harm to a company's reputation. AI's capacity to detect fraudulent activities in real-time is becoming increasingly important for companies seeking to protect themselves from fraudulent activity. The banking and finance sector already has solutions in place that can detect fraudulent transactions by utilizing various types of machine learning algorithms. The application will prevent the transaction from being processed if it detects a potential for fraud and will inform the relevant parties accordingly (Glauner, 2019).

1.3 Artificial Intelligence (AI) in Different Sectors/industries

Artificial Intelligence (AI) has a wide range of applications in business across various industries. The following are some prominent areas where AI is making a significant impact.

1.3.1 Artificial Intelligence (AI) in Healthcare

The contribution of technology giants such as Microsoft, Google, Apple, and IBM to the healthcare industry is of paramount importance. Artificial Intelligence (AI) is being utilized in a range of healthcare applications, such as medical imaging, medication administration, drug research, robotic surgical operations, and data mining for pattern recognition and subsequent more accurate diagnosis and treatment of medical conditions (Bharadiya, Tzenios & Reddy, 2023). IBM Watson, an AI tool, can be utilized to analyse a patient's medical records in order to identify potential treatments. This is achieved by assessing the importance and context of a combination of structured and unstructured data that may have relevance in determining an appropriate treatment plan. Similarly, a platform developed by the biopharma business NuMedii, Artificial Intelligence for Drug Discovery (AIIDD), utilizes big data and AI to identify the correlation between illnesses and medications at the systems level (Poduri, 2021).

Furthermore, AI can be used as a health monitoring tool. Wearable gadgets integrated with AI have the capability to monitor the health of patients and provide timely alerts. Wearable technology pertains to health monitoring devices that are worn on the body, often on the wrist, with the purpose of monitoring and collecting the wearer's health data. These gadgets use biosensors to gather diverse patient data, including heart rate, blood pressure, sleep habits, and activity.

1.3.2 Artificial Intelligence (AI) in Finance and Banking

The development of Artificial Intelligence (AI) applications has had a considerable impact on the banking and financial services industry. There is a vast array of AI applications in use in this sector. In many instances, advanced software robots are supplanting human workers to process loan applications in a matter of milliseconds. Similarly, robotic financial advisers are able to rapidly analyze multiple layers of data to recommend the most suitable investments for their clients (Felzmann, Villaronga, Lutz, & Tamò-Larrieux, 2018). AI is increasingly being utilized in the insurance industry to improve customer experience and create insurance products and solutions that are informed by consumer data. AI-based chatbots have significantly sped up the processing of claims, which has been beneficial to both consumers and insurance providers (Galloway & Swiatek, 2018).

Fraud detection is a key application of AI in the banking industry which can be seen in the example of Mastercard analyzing multiple data points to detect fraudulent transactions through AI-based Decision Intelligence technology (Gregory et al., 2021). AI can identify unusual patterns in transactions to detect and prevent fraudulent activities. In addition to this, artificial intelligence in the form of algorithmic trading powers algorithms that carry out transactions based on real-time market data and predetermined plans.

1.3.3 Artificial Intelligence (AI) in Transportation

The transportation industry is on the verge of undergoing a major transformation driven by Artificial Intelligence (AI). AI algorithms are increasingly being used to support tasks such as the fastest shipping route and last-mile delivery. Self-driving vehicles are expected to be the next major advancement in the transportation sector, with AI-based self-driving vehicles potentially replacing manual driving and improving road safety. Currently, these vehicles are still in the testing and research stages in many countries, with Tesla, Uber, Volvo, and Volkswagen leading the way in this field (Jarrahi, 2018). Additionally, research is being conducted on how AI algorithms can be used to schedule, route, and even operate traffic lights in public transportation. In addition, electric vehicles are being integrated with AI, which has a significant impact on reducing environmental pollution. An example of this is Connect Transit, which utilizes AI-integrated electric buses.

1.3.4 Artificial Intelligence (AI) in Retail

Retail and e-commerce may be the only sectors in which the majority of final consumers can observe the implementation of Artificial Intelligence (AI) in action. In this highly competitive industry, retail businesses are constantly seeking ways to identify trends in customer behavior in order to better align their strategies and gain an edge over their competitors (Chen, Tao, Wang & Chen, 2015). AI-driven automation enables retailers to save money while increasing productivity, precision, and customer satisfaction. AI-enabled technologies, such as chatbots, VPA, and image recognition, are being employed to automate many of the most labor-intensive and mundane tasks previously performed by store personnel. In addition, AI is being utilized to deliver tailored recommendations to customers, enabling them to locate the products they require in a timely and straightforward manner. Additionally, AI-powered analytics can be utilized to identify customer preferences and trends, enabling retailers to adapt their products and services to meet the requirements of their customers. Lastly, AI-driven automation can assist retailers in optimizing their supply chain processes, enabling timely and accurate delivery of products in appropriate quantities.

1.4 The Evolution of Human Resource Management (HRM)

Human Resource Management (HRM) is a discipline with a rich and extensive history, which has its roots in the study of psychology, organizational behavior, and industrial engineering. Over the years, HRM has been used for a variety of purposes within an organization. Initially, it was primarily a compliance-oriented function and was referred to as the personnel department, HRM was primarily responsible for recruitment and the management of employment-related documentation. However, as time has progressed, it has evolved into a crucial determinant in the development of human resources. According to the writers of the book titled "HR from the Outside" the field of HR has seen many waves of development (Ulrich, 2020).

1.4.1 Evolution of Human Resource (HR) Work in Four "Waves"

(a) **Wave 1—Early 1900s**

The present study is centered on the administrative duties of HR personnel, which encompass the formulation of work terms and conditions, provision of HR services, and adherence to regulatory compliance, particularly in the area of payroll. While this administrative facet remains a crucial aspect of HR operations, it is now commonly executed through alternative means, such as technology and outsourcing solutions.

(b) **Wave 2—1970s**

The Wave 2 of human resource management was characterized by a concentration on the development of inventive HR practice domains, including compensation, learning, and sourcing. HR experts operating in these areas initiated communication and collaboration with one another to establish a uniform approach to human resource management. The credibility of HR professionals during this period was derived from their ability to provide "best-practice" HR solutions.

(c) **Wave 3—2000s**

For the past fifteen to two decades, HR has been striving to align its work with the business strategy or objectives of a company. Pursuing this goal, HR professionals have been tasked with evaluating and enhancing talent, culture, and leadership to support a business's strategy. During this period, HR professionals have shifted their focus to strategic HR, which is the process of transforming strategies into human resources priorities in order to meet business objectives.

(d) **Wave 4—2020 and Beyond (Pandemic)**

The advent of COVID-19 has brought about a significant transformation in the function of HRM. The pandemic has caused widespread disruption in organizations worldwide, leading to a shift in the work paradigm. In order to curb the spread of the virus, employees were compelled to work remotely, thereby necessitating a change from traditional face-to-face collaboration to virtual modes of communication such as email and videoconferencing. In this context, HRM has assumed a more comprehensive role, extending beyond the purview of work-related health concerns to encompass the overall well-being of employees. HRM professionals have taken on the role of quasi-counsellors, providing support and guidance to employees in these challenging times (Fig. 1).

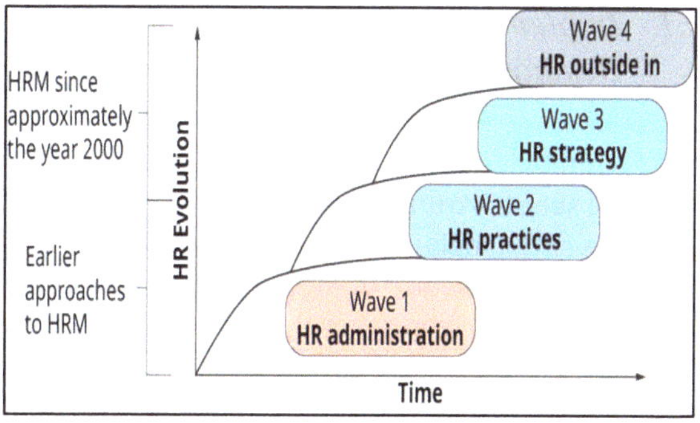

Fig. 1 Evolution of HR work in waves *Source* Bright and Cortes (2019)

2 Reinvention of Human Resource Management (HRM) Practices Through Artificial Intelligence (AI)

The field Human Resource Management (HRM) modernization has experienced an impressive evolution, as digitization integrates into its operational procedures. Since earlier technological inventions such as computers and the internet, HRM effectively used technology to enhance productivity, cost-effectiveness, and market competition (Hmoud & Várallyai, 2020). The integration of Artificial Intelligence (AI) with Human Resource Information Systems (HRIS) is driving the fast advancement of this technology, leading to significant innovation in HR functions. The inclusion of AI in various human resources operations has been prevalent due to the abundance of HR data pertaining to the organization, employees, and tasks that are naturally within HR's responsibility. This integration of AI has been shown to enhance sustainable business models (Di Vaio et al., 2020).

Furthermore, AI-driven human resource analytics has emerged as a popular area within HRM by leveraging data stored in HRIS (Baakeel, 2020). Thus, it enables the redefinition of employee management, particularly in relation to cultivating a professional workforce with the requisite skills and experience necessary for organizational success. The aim is to harness the power and potential of sophisticated AI-enabled systems to guide decision-making. This will enable organizations to develop employees' capabilities, improve teamwork, improve performance management, and support flexible working (Chornous & Gura, 2020).

The field of human resources technology is characterized by rapid advancements. There is a growing trend in utilizing AI to enhance software and hardware solutions that support HR professionals in automating and enhancing various aspects of HRM. These include talent acquisition, recruitment and selection, performance management, payroll, and benefits administration, as well as processes related to re-skilling, and up-skilling (Liu, Ma, Zheng, Liu, & Liu, 2022; Mainka, 2019). Although AI-based solutions are gaining popularity and influence in the field of HRM, it is important to ensure the ethical occurrence of discriminatory and unjust outcomes, as previously acknowledged.

2.1 Implementation of Artificial Intelligence (AI) in Human Resource Management (HRM) Practices

The Fourth Industrial Revolution (IR 4.0) marks an increased use of emerging technologies, such as Artificial intelligence (AI), machine learning, big data, the Internet of Things (IoT), mobile technology, speech recognition, virtual reality, geo-tagging, and biometrics. The use of these sophisticated technologies has revolutionized the manner in which organizations' operations are carried out on a local or global scale, resulting in significant implications for work design, employee engagement, and alterations in workplace procedures. AI refers to the field of science and engineering that focuses on the development of intelligent machines, particularly branches within the broader discipline of computer science, which aims to enhance the capabilities of computers to exhibit intelligent behavior (McCarthy, 1989). AI is an innovative tool used to transform the role of management and organizational practices (Kshetri, 2021). AI has changed the pattern of work and decision-making abilities for many organizations with its smart technological approach such as genetic algorithms, text mining, data mining, sentiment analysis, neural networks, and interactive voice recognition applications (Strohmeier & Piazza, 2015). In addition, it improves the decision-making ability and cost-effectiveness of organizations by making decisions on real-time data.

AI also known as an interdisciplinary science, imitates human capabilities and intellectual behavior. The application of AI technology is also evolving. AI is a new trend in the HR field that has made its mark in the HR field as a recent age. AI has become an important phenomenon in the field of human resources. However, the impact of this phenomenon on employee management cannot be underestimated, as it is widely recognized as an important factor in several aspects such as recruitment and selection, training and development, as well as the growth and retention of employees within the organizational setting (Ivanov & Webster, 2017). Furthermore, the current trend of incorporating machine learning and AI technology into HR practices has promise for enhancing employee engagement and workplace efficiency (Neumann & Bisschops, 2019). Therefore, it is clear that the use of AI facilitates several predictive functions pertaining to recruiting, selection, onboarding, training and development, performance management, compensation and benefits, employee engagement and retention.

2.1.1 Artificial Intelligence (AI) in Recruitment and Selection

The recruitment and selection of highly competent employees is a primary focus for HR departments. The recruitment and selection function is centered on preparation for the future requirements of an organization. The process aligns with the organization's strategy for future endeavours by effectively identifying and selecting the most suitable talents to occupy job positions. The transformations occurring on a global scale have necessitated a paradigm shift for organizations. They are now compelled to modify their conventional recruitment practices and embrace a new method of attracting skilled candidates. Moreover, they must ensure the retention of talent within their organization and remain abreast of the latest advancements in technology. Artificial Intelligence (AI) plays a crucial role in Human Resources Planning (HRP) by analyzing data to forecast future employees' requirements and facilitating the recruiting process. The technology tools of recruitment and selection have been improved with the use of AI and digital technologies such as chatbots, social networks, Massive Online Open Courses (MOOCs), serious games, and big data analytics–matching systems that are complementary forms of digital technologies (Allal-Chérif et al., 2021).

A significant portion of recruiters and human resources professionals devote a substantial amount of their time to the recruiting and selecting process. This process encompasses many tasks such as talent attraction, resume screening, tracking and evaluation, scheduling and performing first interviews, as well as notifying applicants about their application status. The process of recruitment and selection encompasses the posting of job advertisements on appropriate websites and the search for potential candidates, which is traditionally time-consuming. However, advancements in AI have facilitated the integration of machine learning algorithms into popular platforms like Glassdoor, Naukri, LinkedIn, and Indeed. These algorithms leverage various factors such as candidate resumes, keyword usage, search history, and professional connections to generate personalized job recommendations for individuals seeking employment opportunities. Additionally, AI may provide improved communication about job openings. The improvement of job applicants' technological capabilities positively impacts their engagement in the recruiting and selection process.

AI recruitment has become increasingly more efficient at finding and hiring high-quality employees than wholly human-centered recruitment. AI exhibits a diverse array of applications within the functions of HRM, specifically, AI technologies have been used in the recruitment and selection process to enhance the efficiency and effectiveness of job application and selection procedures. The AI helps recruiters with tasks like screening, establishing contact, scheduling meetings and interviews, and engaging with candidates. The process of reviewing applicants' applications and conducting first interviews is a crucial and time-consuming undertaking, particularly when there is a large number of positions to be filled. With the assistance of AI, software businesses have the capability to use video or audio interviews to enhance efficiency and expedite the execution of this activity. These modern technologies possess the ability to swiftly access and analyze large volumes of data, surpassing human capabilities in terms of speed and capacity. Moreover, AI has the potential to improve the process of job applicant identification by effectively determining those who possess the highest level of interest and suitability for a certain position.

AI also allowed HR professionals to identify suitable profiles for job vacancies, eliminating cognitive biases of gender, sexual orientation, and race that involve human judgment in recruiting activities. AI plays a crucial role in enhancing the efficacy of the job interview process. This transformation is evident in the shift from traditional face-to-face interviews to internet-based interviews, namely asynchronous video interviews (AVIs) (Torres & Mejia, 2017). Besides that, according to Pessach,

Singer, Avrahami, Ben-Gal, Shmueli, and Ben-Gal (2020) using a hybrid decision-support tool helped HR professionals in the recruitment and placement processes and increased the impact of recruiters and maximized organizational return on investment.

AI is helping organizations to work efficiently by processing a large number of candidate applications in a short time. Johnson et al. (2020) investigated the integration of AI into HRM processes with the aim of enhancing the recruitment of highly skilled employees, increasing retention rates, and reducing the duration required for employee replacement. The researchers also discovered that AI enables the facilitation of two cognitive factors that contribute to effective decision-making within organizational contexts. The first factor is cognitive insight, which refers to the use of algorithms and machine learning methods by AI to aid in the interpretation of vast amounts of data and discover patterns that were previously unrecognized by the organization. These technological improvements have facilitated organizations in enhancing their operational efficiency, namely in the areas of candidate selection for recruitment and identifying possible internal candidates for promotion within the organization.

The second factor is cognitive engagement, which involves the use of a chatbot to replicate a social interaction that resembles human behavior facilitated by Natural Language Processing (NLP) technology. NLP can convert speech into text within microseconds which increases recruiter's efficiency by eliminating their typing work (Jia, Guo, Li, Li, & Chen, 2018). Bots possess advanced NLP capabilities, enabling them to assume a prominent position in many forms of candidate communications. AI-powered chatbots are used to establish communication with job candidates, address their inquiries, and maintain consistent and engaged interaction with them throughout the recruiting process (Upadhyay & Khandelwal, 2018). Therefore, organizations have the potential to use chatbots, leveraging the underlying technological infrastructure, and evaluating their ability to simulate human-like conversations. There are a diverse range of chatbot and their corresponding architectural designs. The use of AI enables organizations to enhance candidate engagement and establish a consistent and lasting relationship with their candidates.

Besides saving time and cost, other benefits of AI in the recruitment and selection process are hiring quality candidates, mapping talents accurately, reducing biases, and redressing queries of candidates quickly (Geetha et al., 2018). For example, online hiring platforms like HireVue.com are designed to conduct pre-hire assessments. HireVue.com facilitates the use of pre-hire assessment questions by recruiters during selection interviews, whereby applicants are required to respond to these questions using a video format. After that, the pre-recorded videos are assessed by the recruiters to select suitable candidates for a final interview. By using this AI technology in the recruitment and selection process, more applicants can be assessed in a short time. In addition, this process also facilitates the recruiter to engage with candidates and in simplifying the recruiting process.

Many organizations do not effectively engage with applicants due to the time-consuming process. Typically, organizations do not provide feedback or communicate the result of an application or interview to candidates. Job applicants often anticipate receiving a response from the organizations subsequent to submitting their application and participating in an interview session. Therefore, it is crucial to promptly react to their application and conduct an interview, as this may significantly impact the likelihood of retaining them as potential employees. Failure to do so may result in losing them to competitors who demonstrate a high level of responsiveness. AI-integrated software applications, such as Applicant Tracking System (ATS), Chabot, and Customer Relationship Management (CRM), play a crucial role in addressing applicants' inquiries and providing them with real-time feedback on their current status.

2.1.2 Artificial Intelligence (AI) in Onboarding

Onboarding refers to the process of making newly hired employees familiar with the organization's rules and policies, environment, and overall cultural dynamics. Orientation sessions are often conducted by organizations for this purpose. Providing additional support and guidance to newly hired employees is important, but individually managing each of them is a challenging task. Besides that, newly hired employees often encounter challenges while attempting to comprehend the onboarding contents. Artificial Intelligence (AI) is now playing a significant role in facilitating organizational adaptability among newly hired employees by implementing personalized onboarding procedures.

Recently, HR professionals have been receiving support from automated bots designed to engage with recently hired employees. The intelligence bots serve the purpose of providing essential information on employment benefits, regulations, and policies, as well as addressing inquiries and concerns to aid employees in their understanding and resolution. However, most of the employees forget or miss specific information during these orientations. Hence, intelligent bots assist HRs in creating new employee profiles. Besides that, employees also can interact with these chatbots which can be programmed so that employees can clarify company regulations and policies, employee benefits, and insurance in real time.

The acceptance of offer letters and the subsequent onboarding procedure is the last stage of the recruitment process, which has significant importance for the organization. Currently, there is a significant increase in the rates of applicants who drop out after accepting an offer letter. There is a transitional phase, sometimes referred to as the "grey area," that occurs between the submission of a resignation by a candidate in their present employer and their subsequent joining of a new organization. Hence, to maintain their engagement, it is necessary to send frequent follow-ups to decrease the dropout rates, a task facilitated by AI. Employees are more likely to remain with the company if the onboarding process is well-organized and instructive.

2.1.3 Artificial Intelligence (AI) in Training and Development

In the present dynamic context, it is crucial for employees to possess a comprehensive understanding of the most recent trends, advancements, and modifications pertinent to their respective professional development. Hence, the establishment of a suitable training facility is essential in any organization to cultivate a professional and technically skilled employee. Emerging technologies in the field of training and development are aiding human resources professionals in enhancing the efficacy and intelligence of their training programs by using digital and online learning platforms. Besides that, the assessment of employee satisfaction and the evaluation of the benefits derived from training sessions are crucial. Therefore, employees are encouraged to provide feedback and ideas based on their experiences throughout the training period. Artificial Intelligence plays a significant role in supporting the human resource (HR) department by assisting in various feedback-related tasks.

AI supports the training and development of employees more effectively. HR systems save each existing employee's information and provide the organization with an online inventory of its employees. Thus, it helps in the identification of skill deficiencies and the creation of appropriate training initiatives. It can also help organizations in the process of identifying an appropriate candidate within the organization. Virtual and digital training offers many benefits for both the organization and its employees. Employees can use these systems for the purpose of effectively managing their future career prospects. In the event that employees possess deficiencies in their skill sets, AI systems may assist in identifying their training requirements and facilitating the completion of necessary courses. Besides that, AI plays a crucial role in supporting HR managers in evaluating the efficacy of training programs and facilitating decision-making processes related to employee competence. This includes analyzing various aspects such as emotional and intellectual skills, as well as experience levels, in order to appropriately match employees with their respective talents to suitable roles.

Ensuring the continuous professional development and advancement of employees within an organization is crucial for achieving diverse organizational goals. The variability in the number of employees within an organization presents the potential for AI to enhance and optimize training information system procedures, hence improving educational processes. One notable capability of AI is its capacity to provide tailored training recommendations to employees by analyzing their aptitude, interests, and potential for success. These skills enable HR professionals to explore an employee's potential without being overwhelmed by the extensive time and effort required to conduct such analyses for each individual employee. Xu and Xiao (2020) propose the use of AI-augmented virtual reality (VR) simulators as a means to improve mandatory employee training. The authors report a notable increase of 79% in employee involvement as a result of this approach. The authors of this study conducted research to examine the utilization of virtual reality (VR) technology in the context of Human Resource Management (HRM). Their findings highlighted the potential benefits of using virtual reality (VR) technology in HRM practices, including increased efficiency, cost reduction, and enhanced competitiveness for organizations operating in various sectors. Therefore, the use of AI-enhanced training is most effective when employed as an augmentation tools to optimize the process of employee growth. The methods used for enhancement include the monitoring, suggesting, and examination of training activities, affording people a significant role in addressing any deficiencies. Additionally, a customized and individualized approach is incorporated into the onboarding procedure.

The use of AI enables the customization of training and development programmes which provides employees with a more personalised and tailored experience. Therefore, individuals have the opportunity to engage in learning activities according to their own schedules and preferences. HR personnel are responsible for ensuring that the learning material has comprehensive and relevant information, as well as an effective grading system. This approach aims to enhance the employees' concentration on the training material and increase their engagement. The primary objective of HR in the context of AI in training and development is to concentrate on the creation of high-quality training and development programs that align with current trends, emerging skills, and other relevant factors. These programs are tailored to the unique preferences of each individual and are made accessible via an e-learning platform. Therefore, HR managers are not required to allocate their time to the practical coordination and implementation of employee training programs, a task that is often difficult and time-consuming. While online training and development programs provide less human influence, it is crucial to comprehend the employees' perceptions of the sessions they have participated in. Thus, obtaining feedback is crucial as it facilitates the enhancement of training and development initiatives in subsequent periods, while also fostering a sense of appreciation among employees and demonstrating that their suggestions are considered.

2.1.4 Artificial Intelligence (AI) in Performance Management

Performance management is a crucial practice within the field of Human Resource Management (HRM). Performance management systems are specifically developed to assist organizations in effectively managing their workforce by means of establishing objectives, allocating and assessing tasks, and implementing consequences and incentives (Fletcher, 2001). The use of Artificial Intelligence (AI) has significantly expedited and enhanced this particular procedure. Automated processes are now used for the execution of evaluation methods, such as the 360° assessment. The system incorporates employee assessment criteria, among other pertinent data, in order to derive performance-based outcomes.

AI-enabled performance management methods provide various opportunities for both employees and organizations. For example, a fuzzy multi-attribute decision-making tool may result in an equitable assessment of employees. Significantly, this tool facilitates the identification of employees who need more improvement in certain areas and quantifies the extent of changes required. Furthermore, AI performance management tools facilitate the evaluation of employee performance by HR managers and offer recommendations for necessary enhancements. These tools also enable the implementation of corrective measures, guided by expert opinions, such as providing training, enhancing talent, and pursuing additional qualifications as deemed necessary.

Considering the capabilities of AI, it has the ability to effectively and expeditiously identify and evaluate potential candidates for employment, while also aligning the goals of the organization with the performance of its employees. The use of AI in the field of HRM facilitates the process of establishing appropriate goals for managers and ensures that all departments align with the organizational vision. The use of accurate information in employee performance evaluations allows managers to make the right judgments. In the process of goal-setting, employees are assigned predetermined objectives or standards that must be accomplished within a certain timeframe. AI has the capability to provide timely and ongoing performance feedback to both employees and managers, facilitating the identification of areas that need improvement. This feedback mechanism also aids in aligning the goals of employees with the overall objectives of the organization.

A significant component of a performance management system is the process of performance assessment and evaluation, which affects the allocation of incentives and penalties. These measures include both the tasks accomplished by an employee and the attributes that are considered essential for the successful performance of the given work. These systems have faced substantial criticism due to concerns over their validity, reliability, and potential bias. Additionally, some employers have expressed doubts regarding their efficacy. However, it is important to note that performance management is inherently a very individualized procedure. The outcome of HR decisions, such as employee selection and termination, have significant implications for people and societal entities that prioritize fairness and justice. The occurrence of errors in AI is generally limited and may be deemed acceptable because of its extensive effectiveness in handling routine and less significant decision-making tasks. Therefore, when considering the use of AI it is essential to strive for an application that is devoid of any bias or unfavourable effects on any particular set of employees.

In addition, AI has the capability to organize skills mapping, enabling the analysis of employee abilities and competencies. This analysis facilitates the identification of potential avenues for career advancement and professional development within the organizational context. This has the potential to serve as a source of motivation for workers, increasing their likelihood of remaining employed within the organization.

2.1.5 Artificial Intelligence (AI) in Compensation and Benefit

Compensation and benefits are a crucial component of Human Resource Management (HRM) that exhibits a clear correlation with employee performance inside the workplace. The procedure pertains to the determination of employee remuneration in accordance with regulations and policies. The implementation of a compensation and benefits management system in the organization has the potential to improve the performance of both individuals and groups. The use of technology in compensation and benefits systems provides HR professionals with assistance in effectively managing various HR payroll tasks and associated value-added functions. Artificial Intelligence (AI) technologies have the capability to monitor several forms of employee data, including personal information of employees, and changes in personal details such as marital status, dependents, and beneficiaries. AI has the potential to assist organizations in customizing remuneration and benefits packages according to the unique requirements and preferences of individual employees, hence enhancing their appeal and personalization. Furthermore, the gap between the supply and demand of skills, as derived from databases, aids in the determination of an organization's compensation and benefits strategies. AI systems also aid managers and specialists in gathering the most relevant information pertaining to the compensation and benefits systems of employees. These systems play a crucial role in the computation and establishment of wage criteria for employees based on their respective job roles.

In the context of AI in compensation and benefits, it is essential for organizations to overcome ambiguous assertions about fairness and instead adopt more precise measurements in order to ensure equitable practice. Hence, to use AI for the purpose of determining an employee's compensation or benefits package, it is important for the organization to assess whether individuals in similar positions work comparable hours and get equitable remuneration, irrespective of individual demographic factors such as gender, age, and race. Data analytics relating to pay equity is crucial in securing fair and ethical compensation and benefits practices within an HR department. The Human Resources Department not only has to balance the needs of the organization's employees, but it must also ensure budget expenditures accordingly. Allowing an AI to undertake the burden of tedious number-crunching and inferring the unfathomable amount of information collected on its employees enables HR professionals with the time to review the analysis and validate fair practices. Besides that, the use of data analytics in the context of pay equality is of utmost importance in establishing equitable and ethical compensation and benefits policies within an HR department.

The HR department is tasked with the dual responsibility of effectively managing the demands of the organization's workforce while also ensuring the allocation of budgetary resources. The use of AI to handle the tedious task of performing complex calculations and drawing inferences from vast amounts of data pertaining to employees provides HR experts with the opportunity to allocate their time to review the analysis and ensure the implementation of fair practices. The availability of AI technology in compensation and benefits has significantly contributed to expediting the assessment process while enhancing its accuracy and overall quality. These technologies have furthermore played a role in the delineation and preservation of fairness in the allocation of rewards, such as a comprehensive program suite designed to assist pension funds in accurately computing various forms of recompense.

2.1.6 Artificial Intelligence (AI) in Employee Relations

The human relations component pertains to technologies that enhance an organization's capacity to cultivate and sustain efficient interpersonal and intrapersonal relations. In the context of interpersonal relationships, organizational competence refers to the capacity of an entity to establish connections and facilitate the flow of information and ideas among two or more persons via various channels, such as face-to-face interactions, online platforms, or written communication. This competence contributes to the development and maintenance of employee-employer relationships. Moreover, intrapersonal relationships are the capacity of an organization to effectively self-reflect on achievements and failures. These interactions exhibit a high degree of self-reflection and perceptiveness towards the surrounding environment.

According to Votto et al. (2021) the utilization of technology in human relations offers two fundamental elements that contribute to an organization's capacity to efficiently oversee its workforce. The first element pertains to employee performance and satisfaction (EPS), including the manner in which an organization and its managers establish relationships with their employees, comprehend employee diagnostics, and retain valuable employees. This pillar offers a technology means to assess employee performance and conduct feedback sessions, enabling managers to guide employee behavior and establish a personal relationship to address achievements and issues in the workplace. In addition, the implementation of a robust technology infrastructure for efficient and proficient management of EPS activities offers managers and employees more opportunities for communication and professional development within the organizational context.

The second component refers to Discipline Management Systems (DMS), which focuses on the rehabilitation of employee behavior and the procedures involved in termination (Tariq et al., 2016). When it comes to DMS, effectively managing disciplinary matters inside an organization requires a comprehensive of the company's policies and procedures. Nevertheless, managing sensitive discussions and potential dismissals requires the possession of robust inter/intrapersonal skills.

2.1.7 Artificial Intelligence (AI) in Employee Engagement

Organizations today possess enhanced capabilities to forecast the degree of employee engagement by using diverse prediction methodologies facilitated by Artificial Intelligence (AI). The aforementioned technical tools possess the ability to analyze extensive data sets, extract significant outcomes from them, and forecast the current and future levels of employee engagement. The use of face recognition technology enables organizations to detect the emotional state of their employees by analyzing their facial expressions on a given day. This facilitates the organization's comprehension of employee behavior, hence fostering a sense of significance and appreciation among employees. Consequently, this ultimately results in an elevated degree of employee engagement inside the organization.

The objective of organizations extends beyond just employee happiness and satisfaction, including the additional goal of employee engagement. The attainment of employee engagement relies on a holistic approach that incorporates several aspects, starting from the initial recruitment of an employee until they leave from an organization. Employee engagement encompasses several factors that contribute to the overall satisfaction and commitment of employees within an organization. These factors include a clear understanding of job responsibilities, chances for continuous learning and development, appropriate incentives and recognition for performance, effective mechanisms for addressing grievances, and initiatives aimed at promoting employee health and well-being. The use of AI in employee engagement has the potential to provide fair and equitable outcomes. Furthermore, the use of AI also customized learning and growth facilitates the ability to make real-time inquiries about opportunities, performance feedback, and employee benefits. Additionally, the utilization of a virtual assistant aids in comprehending business regulations and policies. This significantly increases the level of employee engagement.

AI-driven solutions have the capability to provide tailored advice to managers about strategies to enhance employee engagement and satisfaction, using individual employee data. This may include recommendations for enhancing one's professional growth, exploring training prospects, or fostering a better equilibrium between work and personal life. AI-powered chatbots and virtual assistants have the capability to provide immediate responses to HR-related inquiries, hence mitigating employee dissatisfaction and enhancing the entire employee experience. Thus, AI has the potential to promote employee engagement and improve employee satisfaction via its ability to provide prompt and real-time responses to employees. Employees are likely to have higher levels of job satisfaction when they are provided with clear guidance and support in terms of their career path and opportunities for professional development.

Al also assists in the implementation of a continuous feedback mechanism that is impartial and grounded in factual evidence. By using AI techniques, HR departments may effectively formulate and establish goals for individual employees. The use of continuous feedback systems facilitates the process and enhances the outcomes, hence leading to improved results and performance. Other significant aspects are open feedback, suggestions, and concerns expressed by a large number of employees via an online survey also can be facilitated through the utilisation of AI. Moreover, AI plays a crucial role in analyzing the vast amount of data generated by these inputs, enabling not only the assessment of current engagement levels but also the prediction of future engagement, turnover rates, performance levels, and other related factors.

2.1.8 Artificial Intelligence (AI) in Employee Retention

The impact of performance feedback supplied by Artificial Intelligence (AI) on employees' productivity was shown to be much more favorable when compared to feedback provided by humans. Furthermore, employee retention is a critical concern for organizations, as high turnover rates can be costly and disruptive. AI is also contributing to the enhancement of staff retention inside organizations. AI as predictive analytics, can analyse huge amounts of data to identify patterns and factors contributing to employee turnover. AI analyses historical data and predicts which employees are at risk of leaving, allowing organizations to take proactive measures. An AI system analyses the data over a period of one month. Any indication that the employee has the potential to quit is immediately reported to the employer, allowing them to take appropriate measures aimed at mitigating such situations. AI-driven sentiment analysis solutions have the capability to effectively track and analyze employee input across various channels, including surveys, emails, and social media postings. This enables organizations to assess the overall sentiment of their employees and proactively identify any underlying concerns that may contribute to employee turnover.

Furthermore, AI has the capability to analyze exit interview data to discover factors contributing to employee leaving. This data may be used to effectively tackle systemic challenges and provide specific enhancements aimed at retaining prospective employees. Simultaneously, AI has the potential to improve workforce planning activities. AI has the potential to assist organizations in forecasting their future employees requirements, therefore enabling them to strategically align their workforce with the organization's goals. This practice has the potential to mitigate excessive workload and burnout, hence positively impacting employee retention rates.

In addition, AI has the capability to provide comprehensive succession planning. AI has the capability to aid in the identification of prospective successors for critical positions within an organization, therefore mitigating the negative impact caused by unforeseen departures. The use of AI has considerable promise for enhancing employee retention; nonetheless, it is essential to utilize these technologies in an ethical and transparent manner. It is essential that employees see the use of their data as advantageous to their own interests rather than as a kind of intrusive monitoring. Furthermore, it is imperative that AI serves as a supplementary tool rather than a substitute for human HR specialists. These professionals possess the essential ability to provide personalized assistance and emotional guidance, which are often crucial in endeavours aimed at retaining employees.

3 Usage of Artificial Intelligence (AI)-Based Software in Human Resource Management (HRM)

The variety of cutting-edge technologies provides robots with the ability to carry out tasks that people undertake across a range of domains, including cognition, perception, and execution (Akerkar, 2019). This development is in line with the change in Human Resource Management (HRM) brought about by the use of Artificial Intelligence (AI)-based software, which has altered the environment by streamlining decision-making procedures, automating routine tasks, and enhancing employee engagement.

3.1 Types of Artificial Intelligence Software Tools Usage for the Human Resource Industry

Artificial intelligence (AI) enhanced tools have revolutionized processes like hiring, performance review, and workforce planning. These tools range from the use of Robotic Process Automation (RPA) that streamlines administrative operations to the exploitation of cloud-based services that enable remote accessibility and real-time data analysis. Furthermore, Data Workbenches give data-supported insights to direct strategic workforce activities, while Microservices provide modular adaptability in designing HR solutions. AI that is integrated into processes improves engagement and recruitment efforts, and AI First Suites offer comprehensive HR services that are powered by AI expertise. The fusion of AI technology enables HR professionals to focus on strategic initiatives, foster employee development, and support the success of the firm as a whole.

3.1.1 Robotic Process Automation (RPA)

The use of AI in Human Resource Management (HRM) is being transformed by Robotic Process Automation (RPA). RPA uses intelligent software robots to imitate human interactions with digital systems, executing rule-based tasks quickly and accurately. The importance of RPA lies in automating a variety of routine processes in HRM, such as employee onboarding, payroll processing, and leave management. The automation of these bots allows them to handle data entry, validate information, and even perform compliance checks, freeing HR professionals from time-consuming administrative tasks.

3.1.2 Cloud-Based Services

Artificial Intelligence (AI) powers Cloud-based Services to revolutionize Human Resource Management (HRM). AI-powered cloud platforms enable HRM to quickly and efficiently access and analyze massive amounts of data. AI algorithms are employed in these platforms to perform predictive analytics, enabling HR professionals to anticipate workforce trends and make informed decisions regarding talent acquisition, retention, and succession planning. Additionally, cloud services that are AI-enhanced enable seamless collaboration among HR teams, regardless of their geographical location.

3.1.3 Data Workbench

The Data Workbench, which is infused with Artificial Intelligence (AI), has become a crucial tool for Human Resource Management (HRM). The power of AI-driven data analysis enables HRM to extract valuable information from complex datasets. AI algorithms in Data Workbenches enable predictive modeling and trend identification, which aids HR professionals in making informed decisions about workforce planning, performance assessment, and talent development. Historical and real-time data analysis allows AI-equipped Data Workbenches to forecast attrition rates, identify skill gaps, and recommend strategies for employee engagement. By detecting potential bias in HR processes and suggesting corrective actions, this technology facilitates diversity and inclusion efforts. At the end of the day, AI data workbenches enable HRM to take a proactive and strategic approach, transforming raw data into actionable data that drives organizational growth and human capital development.

3.1.4 Microservices

Microservices, integrated with artificial intelligence (AI) capabilities, are reshaping Human Resource Management (HRM) practices. These modular units and independently deployable application capabilities are improved with AI algorithms to simplify various HR processes. By automating services like benefits enrolment, performance tracking, and learning programs in accordance with user needs and preferences, AI-powered microservices offer tailored employee experiences. AI in Microservices can provide specific training programs, mentoring opportunities, and career promotion paths by examining employee data and patterns. Employee engagement, productivity, and retention are improved by this innovative method. The alignment of HR strategy with current market developments and organizational objectives further enables adaptive workforce planning through AI-infused microservices. It is possible for HRM to adopt a more flexible, quick-to-respond, and employee-centric strategy through the use of AI and Microservices, which will improve workforce management and promote a culture of continuous improvement.

3.1.5 Embedded Artificial Intelligence (AI)

In Human Resource Management (HRM), embedded Artificial Intelligence (AI) is becoming a transformative force, transforming traditional practices by seamlessly integrating intelligence into processes and systems. The integration of this AI into HR tools and platforms is discreet, enhancing functionality and providing user experiences that are intuitive. By intelligently matching candidate profiles with job criteria, embedded AI improves talent acquisition in HRM, speeding up the hiring process and increasing candidate fit. It also enhances employee engagement by assessing the attitudes expressed in messages and interactions, enabling HR managers to proactively solve issues and raise workplace happiness. Embedded AI can forecast attrition risks and suggest customized retention tactics through real-time data analysis, which improves workforce management. This integration of AI with current HR systems equips workers to make data-driven decisions, increase efficiency, and promote an innovative culture within the company.

3.1.6 Artificial Intelligence (AI) First Suite

Embedded AI has become a transformative force in Human Resource Management (HRM), transforming traditional practices by seamlessly integrating intelligence into processes and systems. The subtle integration of this AI into various HR tools and platforms enhances functionality and provides intuitive user experiences. By intelligently matching candidate profiles with job criteria, embedded AI improves talent acquisition in HRM, speeding up the hiring process and increasing candidate fit. Additionally, it improves workplace satisfaction by assessing attitudes in messages and interactions, which enables HR professionals to proactively address issues. Embedded AI can forecast attrition risks through real-time data analysis and provide customized retention measures, which improves labor management. Professionals are given the ability to make data-driven decisions, increase productivity, and promote an innovative culture within the company thanks to the integration of AI into existing HR systems.

4 Opportunities and Challenges in a Digital World of Work Associated with Artificial Intelligence (AI) in Human Resource Management (HRM)

The adaptation of information and communication technologies (ICT) in the Human Resource Management (HRM) practices is not a new phenomenon. For instance, the use of Human Resource Information System (HRIS) applications and some other software developed for managing routine and non-routine HRM activities in organizations since the mid-1980s (Budhwar et al., 2022). However, the existence of AI serves a variety of cutting-edge technologies across an extent of domains such as cognition, perception, and execution (Akerkar, 2019). Thus, there are opportunities and challenges associated the AI in HRM practices at organizations.

4.1 Opportunities of Using Artificial Intelligence (AI) in Human Resource Management

The use of Artificial Intelligence (AI) in Human Resource Management (HRM) practices will create a way forward for better management of an organization (Arslan et al., 2022). HRM serves an important role in the organization by producing effective and productive human capital. However, the existence of AI could boost the HRM functions to enhance the human capital in the organization. According to Kiron and Spindel (2019), most HR leaders or managers look forward to AI in their organization specifically for a wide range of tasks such as recruitment, training, onboarding, and employee performance records. Furthermore, AI can handle repetitive functions and

HR professionals can focus on creating strategies and improvements for the human capital (Vrontis et al., 2022). Thus, both AI and HR professionals can work together, and the result will be beyond the expectation. Hence, the existence of AI in HRM practices will add more valuable insight and drastically improve efficiency for the HR departments in the organization.

4.1.1 Recruiting and Examining of Applications

As we know Artificial Intelligence (AI) is a simulation of human intelligence in machines that can make decisions as human beings by using algorithms. The literature on AI focuses more on the type of intelligence of the application used in the organization. The application could replace the Human Resource Management (HRM) practices in a good way and efficiency. According to Mitrofanova and Konovalova (2019), AI can replace routine tasks that could help HR professionals in doing it effectively. Moreover, HR professionals can concentrate on non-routine tasks and strategies for the future of HRM. As the AI in HRM, it can help in certain functions such as recruitment and selection, performance management, training, and compensation. AI can be applied in the recruitment and selection process by scanning resumes, cover letters, and even the background of candidates by scanning the candidates' social media profiles. This could be easier when using AI compared to the HR professional (human) because this task is considered a routine activity for the recruitment team. However, HR professionals still can monitor the AI selection process for added value. This AI-based be able to support data-driven decision-making by analysing and deriving astute knowledge from multiple streams of heterogeneous big data.

Moreover, AI has the capability to efficiently navigate extensive databases and social media platforms in order to discern prospective candidates. The system has the capability to analyze resumes and profiles in order to identify and choose the most relevant applicants. AI is capable of analyzing previous recruiting data in order to make predictions on the likelihood of success for applicants in certain jobs. The identification of variables that contribute to employee turnover may assist in the improvement of applicant selection processes. Applicant tracking systems powered by artificial intelligence have the potential to enhance the efficiency of the recruiting process via the use of advanced algorithms to align candidate abilities with job prerequisites. Chatbots have the capability to provide candidates with relevant information and offer guidance during the application procedure. The use of automated onboarding procedures has the potential to reduce administrative workload and provide a seamless transition for newly recruited employees.

Furthermore, AI-enabled recruiting and selection processes are important in the acquisition and selection of highly skilled employees for organizations. The use of AI in recruiting processes has been shown to reduce the duration of the recruitment cycle, allowing businesses to promptly address emerging circumstances and eventually enhance their competitive edge by using intangible assets associated with the acquisition of highly qualified employees. Thus AI provides many possible benefits, including the ability to rapidly scan resumes, automatically react to applicant inquiries, and facilitate virtual recruitment operations.

4.1.2 Efficient Onboarding

AI-powered onboarding will be the most important practice that can ensure employees will perform well in the organization. This can be considered as the learning foundation for the new human capital when they join the organization. Employee onboarding is a critical process for any organization both public and private industries. The new employees must be introduced to the culture, policies, values, and norms of the organization. The using of AI in the onboarding process could help HR professionals do this task efficiently because it is a routine practice. There are several ways to automate employee onboarding using AI applications such as creating employee profiles, tracking employee progress, creating and delivering learning materials for both new and existing employees and the online or in-house training program (Verma & Bandi, 2019).

4.1.3 Learning and Development

AI-excellent for training and Development are highly connected which can help the organization to assess the training gaps analysis, cognitive and skill gaps, and develop a better curriculum for training purposes. Furthermore, AI could run an online in-house training program for the employees, and it could be a new approach if it is successful. For instance, an organization can split the training and development program into two approaches: physical and virtual training (AI-based application). Thus, the HR professionals only focus on the physical training, and for the AI-based training, the employees can go through at any time. Hence, AI-based virtual training can offer real-time assistance to employees during their training period.

4.1.4 Employee Engagement and Retention

The concept of succession planning has significant importance for organizations as it facilitates the seamless transfer of leadership and sustains uninterrupted company operations. Artificial Intelligence (AI) has the capability to recognize probable successors for important positions within an organization. The use of AI has the potential to greatly augment the effectiveness of succession planning via the provision of data-driven insights and the facilitation of a more strategic and objective approach. AI also has the capability to examine previous performance data, abilities, competencies, and career growth in order to discern individuals with a high potential for success inside an organization. Machine learning algorithms have the potential to assist HR teams in identifying people who have a higher probability of achieving success in leadership positions.

Moreover, AI has the capability to forecast an employee's future performance by analyzing previous data and evaluating their present skills. This facilitates the identification of people who possess both competence and a high likelihood of excelling in leadership positions for organizations. AI offers impartial perspectives, hence mitigating the impact of biases in the process of making succession choices. HR teams may enhance their decision-making process by using data, which enables them to effectively identify and choose suitable individuals for succession planning purposes.

4.2 *Challenges of Using Artificial Intelligence (AI) in Human Resource Management (HRM)*

While Artificial Intelligence (AI) has the potential to significantly benefit Human Resource Management (HRM), there are several challenges and concerns associated with its implementation in this field. To overcome these challenges, organizations should adopt a cautious and ethical approach to implementing AI in HRM. This includes compliance with relevant laws and regulations, ensuring fairness, and as well as addressing concerns related to data privacy and security. Additionally, organizations should engage in ongoing employee education and training to build trust and acceptance of AI-driven HR processes.

4.2.1 Artificial Intelligence (AI) Biases in Recruitment

Recruitment is one of the important functions to ensure that the organization hires talented individuals for the right position. The adoption of Artificial Intelligence (AI) in recruitment could be biased because of the historical data used to train the algorithms. It is occurring when algorithms are trained on data and if human judgment is captured in the historical record, then there is a very stake that bias will be hallowed in the algorithm itself. For instance, bias can also be established in how the algorithm is intended, and how humans interpret AI outputs. In addition, the algorithm may be programmed to have a preference for certain candidates who have certain skills, knowledge, or experience, which could inadvertently discriminate against certain candidates. Hence, it could lead to employee displacement and close the opportunities for the other group of candidates.

4.2.2 Artificial Intelligence (AI) Lacks Human Touch

Human touch is a key to contextual understanding and monitoring of Human Resource Management (HRM) practices. As HR professionals, enhancing corporate communication is a must in handling human capital. However, the existence of Artificial Intelligence (AI) has forecasted that machines will surpass humans in many complex tasks soon. According to Kshetri (2020), AI refers to cognitive intelligence which definitely can replace the HR professional's task. In addition, if the organization used to apply AI in HRM, the only attachments of HR professionals are on non-routine HR practices. Thus, it will create a lack of human touch in conducting the HRM practices.

Next, AI in HRM will lessen the emotional intelligence that is essential for HR professionals. Emotional intelligence enables HR professionals to understand employee concerns and support the organization (Prentice et al., 2020). AI-based applications cannot empathize with employees or interpret nonverbal cues precisely. Therefore, HR professionals must be involved in at least 50% of the HRM practices in the organization. Thus, the AI applications cannot be maximized and used in the organization without presenting the HR professionals. Hence, HR professionals play a vital role in creating a positive vibe in the working environment and increasing employee engagement.

4.2.3 Increased Risks to Cybersecurity

The Artificial Intelligence (AI)-based approach should heavily rely on vast amounts of data, including personal and sensitive information about employees in the organization. The risk of data and how the organization could protect it still become crucial. Cybersecurity risk is the probability of exposure or loss resulting from a cyber-attack or data breach on your organization. Chatbots can be great vehicles for streamlining routine HR transactions.

The safeguarding of HR data raises many concerns for the advancement of AI in work environments. The prevalence of cyber-related events has led to the expansion of cybersecurity beyond the confines of computer systems, extending its reach to other departments inside organizations. Notably, the HR department has emerged as a crucial component of the corporate defense against cyber threats, serving as the primary gateway for potential security breaches. Data privacy breaches are a common occurrence in AI-powered HR systems due to their reliance on data that may include Personally Identifiable Information (PII) of employees. The potential misuse of employees' Personally Identifiable Information (PII) by hackers presents a significant threat to their safety and well-being. Advancements in data breach technologies are concurrently enhancing, making attacks more unexpected.

4.2.4 Introducing Machine-Generated Errors

The computer is not always the right choice for doing an analysis and decision-making. There will be an error that can lead to the misunderstanding of data interpretation and decision-making considerations (Votto et al., 2021). The AI only considers the information that has been set early and cannot for beyond demands. There are some errors related to AI-based applications that can lead the misunderstanding in organization such as unclear goals, incorrect implementation strategy, and biased data (Sharma, 2021). Firstly, unclear goals are one of the most common errors when the organization implements AI in its practices. The HR professionals must ink the organizational goal and the algorithms of AI applications. Before AI is set up, the organization must have a whole idea to build the AI significantly with the aim of the organization. Hence, the misunderstanding can be avoided while the system is already on track. The organizational vision and mission are a must to achieve and that is why HRM department should be really concerned about using AI. Next, sometimes AI implementation could fail because of the wrong strategies applied. This may seem obvious from poorly structured datasets by the AI. These issues arise because lack of thorough understanding of how AI works or unclear objectives for implementation.

4.2.5 Required an Improved Skill Set

The use of Artificial Intelligence (AI) in Human Resource Management (HRM) is still being developed and is far from being perfect. The technology moving so fast, and the skillset must be up to date. Not everything that AI technologies give is entirely accurate. Some data that have been slotted in the AI is not the latest version and needs to be updated from time to time. Thus, it can create a bias to the candidates if it is about the recruitment and selection process. Furthermore, by having AI in HRM, the HR professionals still have strategies for updating the skillset of the AI used (Votto et al., 2021). The improvement of skillset is required by a professional on that technology and not everyone is ready to do that. Hence, upgrading the AI skillset will cost money and time and the organization must be ready for that. According to Abdeldayem and Aldulaimi (2020), if the HR professionals are not ready for it, AI technologies cannot be implemented yet because they can mess up the organization. The changes do not happen instantly because the organization must be ready for new challenges while using AI technologies. Decades may pass until all the big or small businesses decide to switch to AI for their HRM practices. Hence, the organization must run the survey first before really implementing this AI in the organization.

4.2.6 . Ethical Issue

The use of Artificial Intelligence (AI) in Human Resource Management (HRM) practices must be in an ethical way to make sure that the employees are aware of the data and information used by the AI. According to Moosajee (2019), the use of AI in HRM could be difficult such as several racist and sexist biases are prevalent in some AI applications. For instance, all employees must know how AI-based outcomes affect their performance, attitudes, and behavior in an organization (Connelly et al., 2021). In addition, the employees must recognize the data extraction and data maintenance which is a bit tricky. However, the organization should collaborate on the ethical use of AI in HRM especially in decision-making. According to Rodgers et al. (2023), ethical guidelines or frameworks must be developed if the organization would like to apply AI in its HRM practices. This is because the guidelines and framework can ensure fairness, transparency, accountability, and integrity in HRM practices. Thus, the ethical guidelines and framework can attract employees' trust in the system and approach.

5 Benefits of Artificial Intelligence (AI)

The field of artificial intelligence (AI) is experiencing rapid growth and holds the potential to revolutionize various industries and enhance numerous facets of our existence. The following are the benefits of AI for HR professionals, employees, and organizations.

5.1 HR Professionals

Artificial intelligence (AI) has emerged as an indispensable component of the toolsets employed by HR professionals, and for valid reasons. The utilization of AI-powered tools and technologies enables the execution of tasks that would have otherwise required several hours or even days to accomplish manually. Consequently, recruiters and HR professionals can now devote their attention to high-value, strategic tasks that bolster the overall talent management and development strategies of their respective organizations. AI has been observed to have a positive impact on HR by enabling companies to achieve significant savings in terms of time and financial resources. According to a survey conducted among HR professionals, a majority of 92% acknowledged that AI has facilitated time savings in administrative tasks (Fanning, 2019). This is attributed to the ability of AI-powered software to streamline and automate HR processes, thereby reducing the workload of HR teams and enhancing their productivity levels. The utilization of AI in the recruitment process has been demonstrated to yield significant benefits. For instance, AI can be employed to screen resumes and shortlist candidates, resulting in a reduction of up to 60% in the time required to fill a vacant position (Dhadda, 2023). Additionally, AI can facilitate the scheduling of interviews and communication with candidates, thereby reducing the workload of human resources personnel who would otherwise be required to engage in individualized email or phone correspondence with each candidate.

5.2 Employees Experience

Artificial Intelligence (AI) possesses the necessary capabilities to provide valuable insights into the interactions among employees and how to optimize their potential as a cohesive unit. By leveraging AI, one can obtain critical information regarding the most effective team compositions and identify employees who are best suited to lead specific projects. Such data would be arduous and time-consuming to acquire without the aid of AI, but its utilization can significantly enhance the operational efficiency and quality of work of a team. Furthermore, AI can serve as a valuable tool for addressing mental health concerns in the workplace (Denecke, Alrazaq & Househ, 2021). AI has the capability to generate a customized plan for assisting an employee who is experiencing mental health issues. Additionally, it can suggest effective techniques for enhancing mental well-being within the workplace.

The utilization of AI in the workplace has been found to enhance individuals' sense of autonomy by facilitating their ability to learn from past actions, anticipate the outcome of current actions, and receive feedback that can aid in performance improvement (Xu, Xue & Zhao, 2023). The implementation of AI technology has also been observed to reduce the need for extensive management oversight, thereby fostering a greater sense of autonomy among workers and potentially enhancing job satisfaction. The significance of relationships in fostering workplace contentment has been well-established through AI. Research has indicated that the utilization of AI tools can facilitate the development of robust relationships with colleagues, supervisors, and individuals from other departments by enhancing communication and promoting cooperation. Furthermore, it has been observed that the use of AI tools can enhance the efficacy and reliability of workers, thereby strengthening their relationships with customers.

5.3 Organization

The advent of Artificial Intelligence (AI) automation has brought about a significant transformation in the operational landscape of organizations (Tuomi, 2019). By harnessing the potential of AI technologies, organizations can optimize their processes and automate tasks to enhance efficiency and minimize costs. This technology can facilitate the reduction of manual labor, enhance precision, and optimize workflow. Additionally, it can automate routine tasks such as data entry, customer service inquiries, and marketing campaigns. Furthermore, artificial intelligence automation can create customized automated workflows that cater to the unique requirements of each business. With the appropriate tools, organizations can leverage AI automation to streamline their processes and automate tasks to achieve optimal efficiency. The utilization of AI-powered collaboration tools can furnish teams with valuable insights into their dynamics, thereby enabling them to pinpoint areas that require improvement and make informed decisions. Through the exploitation of

AI's capabilities, organizations can enhance their communication and collaboration within the workplace, ultimately leading to heightened productivity. The utilization of artificial intelligence technology has the potential to enhance operational efficiency, minimize expenses, and enhance customer service. The integration of AI in organizational settings can provide a strategic advantage, enabling organizations to remain at the forefront of a constantly evolving market.

6 Conclusion

The approach of using Artificial Intelligence (AI) in Human Resource Management (HRM) has been discussed in this chapter by looking at the need for AI and its impacts on the organization. As AI becomes increasingly integrated into HR processes, it is vital to navigate the ethical considerations and the continuity of using it in the organization. However, HR professionals must be proactive in handling and addressing issues related to AI-based approaches. It is necessary to maintain the ethical standards and the effectiveness of the application. The organization can incorporate the latest AI applications in their decision-making process even in the most complex situations. The existence of AI in HRM could help the organization move to a competitive advantage in their industry. Despite this, organizations must accept the challenges while using AI in organizations because of the current demand. In fact, to stay in the competitive advantage, the organization must take some risk in using this AI in HRM.

The proper integration of AI into human resources (HR) practices necessitates a lengthy journey, wherein diligent effort is needed to achieve success. One of the primary concerns confronting the prospective workforce pertains to the accessibility of skills necessary to effectively adapt to rapid technological advancements. It is essential to guarantee that the labor force has the requisite skills to effectively support and facilitate the implementation and use of emerging technologies. The advent of digital technology has resulted in a growing divide between industrialized and developing nations, with the distance between them expanding dramatically. The severity of the situation is not solely attributed to decisions regarding the cost of these technologies, but also to the inherent complexity of the high-level technical and professional skills necessary for the design, operation, and maintenance of digital infrastructure. Additionally, there is a pressing need to acquire fundamental skills and achieve proficiency in information and communication technology. The underlying proposition posits that the acquisition and cultivation of skills have significant importance in mitigating inequality and bridging the knowledge disparity prevalent within the workforce. The duty of comprehending the present state of AI in HR processes accompanies the development and advancement in capabilities. Therefore, this necessitates HR practitioners and academics to engage in a thorough examination of existing literature that illuminates the augmented HR capabilities facilitated by AI and identifies areas of potential growth within the HR field.

References

Abdeldayem, M. M., & Aldulaimi, S. H. (2020). Trends and opportunities of artificial intelligence in human resource management: Aspirations for public sector in Bahrain. *International Journal of Scientific and Technology Research, 9*(1), 3867–3871.

Akerkar, R. (2019). *Artificial intelligence for business*. Springer.

Allal-Chérif, O., Aranega, A. Y., & Sánchez, R. C. (2021). Intelligent recruitment: How to identify, select, and retain talents from around the world using artificial intelligence. *Technological Forecasting and Social Change, 169*, 120822.

Arslan, A., Cooper, C., Khan, Z., Golgeci, I., & Ali, I. (2022). Artificial intelligence and human workers interaction at team level: A conceptual assessment of the challenges and potential HRM strategies. *International Journal of Manpower, 43*(1), 75–88.

Baakeel, O. A. (2020). The association between the effectiveness of human resource management functions and the use of artificial intelligence. *International Journal of Advanced Trends in Computer Science and Engineering, 9*(1), 512–606.

Bharadiya, J. P., Tzenios, N. T., & Reddy, M. (2023). Forecasting of crop yield using remote sensing data, agrarian factors and machine learning approaches. *Journal of Engineering Research and Reports, 24*(12), 29–44.

Budhwar, P., Malik, A., De Silva, M. T., & Thevisuthan, P. (2022). Artificial intelligence–challenges and opportunities for international HRM: A review and research agenda. *The International Journal of Human Resource Management, 33*(6), 1065–1097.

Chen, J., Tao, Y., Wang, H., & Chen, T. (2015). Big data based fraud risk management at Alibaba. *The Journal of Finance and Data Science, 1*(1), 1–10.

Chornous, G. O., & Gura, V. L. (2020). Integration of information systems for predictive workforce analytics: Models, synergy, security of entrepreneurship. *European Journal of Sustainable Development, 9*(1), 83–83.

Connelly, C. E., Fieseler, C., Černe, M., Giessner, S. R., & Wong, S. I. (2021). Working in the digitized economy: HRM theory & practice. *Human Resource Management Review, 31*(1), 100762.

Bright, D. S., & Cortes, A. H. (2019). An introduction to human resource management. Principles of management. *OpenStax*. https://openstax.org/books/principles-management/pages/11-1-an-introduction-to-human-resource-management

Denecke, K., Abd-Alrazaq, A., & Househ, M. (2021). Artificial intelligence for chatbots in mental health: opportunities and challenges. In *Multiple perspectives on artificial intelligence in healthcare: Opportunities and challenges* (pp. 115–128).

Dhadda, R. (2023). How AI and machine learning is changing CV fact checking. *The HR Director*. https://www.thehrdirector.com/how-artificial-intelligence-and-machine-learning-can-help-in-resume-screening/

Di Vaio, A., Palladino, R., Hassan, R., & Escobar, O. (2020). Artificial intelligence and business models in the sustainable development goals perspective: A systematic literature review. *Journal of Business Research, 121*, 283–314.

Fanning, K. (2019). The impact of AI on human resources, as told by HR departments. *Online Marketplaces*. https://www.onlinemarketplaces.com/articles/the-impact-of-ai-on-human-resources-as-told-by-hr-departments/

Felzmann, H., Villaronga, E. F., Lutz, C., & Tamò-Larrieux, A. (2018, November). How transparent is your AI? Ethical legal and societal issues of the transparency principle in cyber-physical systems. In *Proceedings of Amsterdam Privacy Conference* (pp. 1–29).

Fletcher, C. (2001). Performance appraisal and management: The developing research agenda. *Journal of Occupational and Organizational Psychology, 74*(4), 473–487.

Galloway, C., & Swiatek, L. (2018). Public relations and artificial intelligence: It's not (just) about robots. *Public Relations Review, 44*(5), 734–740.

Geetha, R., & Bhanu, S. R. D. (2018). Recruitment through artificial intelligence: A conceptual study. *International Journal of Mechanical Engineering and Technology, 9*(7), 63–70.

Glauner, P. (2019). *Artificial intelligence for the detection of electricity theft and irregular power usage in emerging markets*. Doctoral dissertation, University of Luxembourg, Luxembourg.

Gregory, R. W., Henfridsson, O., Kaganer, E., & Kyriakou, H. (2021). The role of artificial intelligence and data network effects for creating user value. *Academy of Management Review, 46*(3), 534–551.

Hmoud, B., & Laszlo, V. (2019). Will artificial intelligence take over human resources recruitment and selection. *Network Intelligence Studies, 7*(13), 21–30.

Hmoud, B. I., & Várallyai, L. (2020). Artificial intelligence in human resources information systems: Investigating its trust and adoption determinants. *International Journal of Engineering and Management Sciences, 5*(1), 749–765.

Hughes, C., Robert, L., Frady, K., & Arroyos, A. (2019). Artificial intelligence, employee engagement, fairness, and job outcomes. In *Managing technology and middle-and low-skilled employees: Advances for economic regeneration* (pp. 61–68). Emerald Publishing Ltd.

Ivanov, S., & Webster, C. (2017). Adoption of robots, artificial intelligence and service automation by travel, tourism and hospitality companıes—A cost-benefit analysıs. In *International Scientific Conference "Contemporary Tourism–Traditions and Innovations* (pp. 168–177).

Jarrahi, M. H. (2018). Artificial intelligence and the future of work: Human-AI symbiosis in organizational decision making. *Business Horizons, 61*(4), 577–586.

Jatobá, M., Santos, J., Gutierriz, I., Moscon, D., Fernandes, P. O., & Teixeira, J. P. (2019). Evolution of artificial intelligence research in human resources. *Procedia Computer Science, 164*, 137–142.

Jia, Q., Guo, Y., Li, R., Li, Y., & Chen, Y. (2018). A conceptual artificial intelligence application framework in human resource management.

Johnson, R. D., Stone, D. L., & Lukaszewski, K. M. (2020). The benefits of eHRM and AI for talent acquisition. *Journal of Tourism Futures, 7*(1), 40–52.

Kaplan, A., & Haenlein, M. (2019). Siri, Siri, in my hand: Who's the fairest in the land? On the interpretations, illustrations, and implications of artificial intelligence. *Business Horizons, 62*(1), 15–25.

Kiron, D., & Spindel, B. (2019). Rebooting work for a digital era. *MIT Sloan Management Review*, 0_1–10.

Kshetri, N. (2020). Artificial intelligence in human resource management in the Global South.

Kshetri, N. (2021). Evolving uses of artificial intelligence in human resource management in emerging economies in the global South: Some preliminary evidence. *Management Research Review, 44*(7), 970–990.

Liu, Z., Ma, Y., Zheng, H., Liu, D., & Liu, J. (2022). Human resource recommendation algorithm based on improved frequent itemset mining. *Future Generation Computer Systems, 126*, 284–288.

Mainka, S. M. (2019). Algorithm-based recruiting technology in the workplace. *Texas A&M J. Prop. l., 5*, 801.

McCarthy, J. (1989). *Artificial intelligence, logic and formalizing common sense, philosophical logic and artificial intelligence* (R. Thomason, Ed.).

Mikalef, P., & Gupta, M. (2021). Artificial intelligence capability: Conceptualization, measurement calibration, and empirical study on its impact on organizational creativity and firm performance. *Information & Management, 58*(3), 103434.

Mitrofanova, A. E., & Konovalova, V. G. (2019). Opportunities, problems and limitations of digital transformation of HR management. In *European Proceedings of Social and Behavioural Sciences*.

Moosajee, N. (2019, March 14). *Fix AI's racist, sexist bias*. https://mg.co.za/article/2019-03-14-fix-ais-racist-sexist-bias/

Neumann, H., & Bisschops, R. (2019). Artificial intelligence and the future of endoscopy. *Digestive Endoscopy, 31*(4), 389–390.

Pessach, D., Singer, G., Avrahami, D., Ben-Gal, H. C., Shmueli, E., & Ben-Gal, I. (2020). Employees recruitment: A prescriptive analytics approach via machine learning and mathematical programming. *Decision Support Systems, 134*, 113290.

Prentice, C., Dominique Lopes, S., & Wang, X. (2020). Emotional intelligence or artificial intelligence–an employee perspective. *Journal of Hospitality Marketing & Management, 29*(4), 377–403.

Poduri, R. (2021). Historical perspective of drug discovery and development. In *Drug discovery and development: From targets and molecules to medicines* (pp. 1–10).

Rodgers, W., Murray, J. M., Stefanidis, A., Degbey, W. Y., & Tarba, S. Y. (2023). An artificial intelligence algorithmic approach to ethical decision-making in human resource management processes. *Human Resource Management Review, 33*(1), 100925.

Sharma, G. (2021, November). A literature review on application of Artificial Intelligence in Human Resource Management and its practices in current organizational scenario. In *2021 Fifth International Conference on I-SMAC (IoT in Social, Mobile, Analytics and Cloud)(I-SMAC)* (pp. 594–600). IEEE.

Strohmeier, S., & Piazza, F. (2015). Artificial intelligence techniques in human resource management—A conceptual exploration. *Intelligent Techniques in Engineering Management: Theory and Applications*, 149–172.

Tang, R., De Donato, L., Besinović, N., Flammini, F., Goverde, R. M., Lin, Z., & Wang, Z. (2022). A literature review of Artificial Intelligence applications in railway systems. *Transportation Research Part C: Emerging Technologies, 140*, 103679.

Tariq, O., Sang, J., & Gulzar, K. (2016). Design and implementation of human resource information systems based on MVC: A case study vocational education in Iraq. *International Journal of u-and e-Service, Science and Technology, 9*(11), 15–26.

Torres, E. N., & Mejia, C. (2017). Asynchronous video interviews in the hospitality industry: Considerations for virtual employee selection. *International Journal of Hospitality Management, 61*, 4–13.

Tuomi, I. (2019). *The impact of artificial intelligence on learning, teaching, and education: policies for the future.* JRC Science for Policy Report. European Commission.

Upadhyay, A. K., & Khandelwal, K. (2018). Applying artificial intelligence: Implications for recruitment. *Strategic HR Review, 17*(5), 255–258.

Ulrich, D. (2020). HR's ever-emerging contribution. *Strategic HR Review, 19*(6), 251–257.

Verma, R., & Bandi, S. (2019). Artificial intelligence & human resource management in Indian IT sector. In *Proceedings of 10th International Conference on Digital Strategies for Organizational Success.*

Votto, A. M., Valecha, R., Najafirad, P., & Rao, H. R. (2021). Artificial intelligence in tactical human resource management: A systematic literature review. *International Journal of Information Management Data Insights, 1*(2), 100047.

Vrignat, P., Kratz, F., & Avila, M. (2022). Sustainable manufacturing, maintenance policies, prognostics and health management: A literature review. *Reliability Engineering & System Safety, 218*, 108140.

Vrontis, D., Christofi, M., Pereira, V., Tarba, S., Makrides, A., & Trichina, E. (2022). Artificial intelligence, robotics, advanced technologies and human resource management: A systematic review. *The International Journal of Human Resource Management, 33*(6), 1237–1266.

Xu, D., & Xiao, X. (2020). Influence of the development of VR technology on enterprise human resource management in the era of artificial intelligence. *IEEE Access.* https://doi.org/10.1109/ACCESS.2020.3020622

Xu, G., Xue, M., & Zhao, J. (2023). The relationship of artificial intelligence opportunity perception and employee workplace well-being: A moderated mediation model. *Int J Environ Res Public Health. 20*(3).

Should Human Resources Management "Go Green"?: The Impact of Green Human Resources Management on Employees' Green Behavior, Affective Commitment and Company Green Performance

Guida Dias, Neuza Ribeiro, Daniel Roque Gomes, Alexandra Leandro, and Maria João Santos

Abstract Environmental concerns reflect a crucial theme in the twenty-first century due to the resource scarcity, the prevalence of pollution, among many other aspects. Organisation have played fundamental roles in contributing towards reducing the impact of their actions on the surrounding environment. In this context, green human resource management (GHRM) constitutes a set of sustainable practices carried out within the framework of human resource management. The objective of this study is to verify the impact of GHRM practices on the behaviours and commitment of members of staff as well as on the environmental performance not only of the members of staff but also the organisation. The sample is made up of 212 participants belonging to the human resource departments of different organisations. The results demonstrate that GHRM practices positively affect green behaviours, the environmental performance and the affective commitment. Hence, companies deploying GHRM practices manage to influence the behaviours of their staff, making them more environmentally aware and, consequently, helping them to improve the

G. Dias · N. Ribeiro
Center of Applied Research on Management and Economics—CARME, School of Technology and Management, Polytechnic Institute of Leiria, Leiria, Portugal

D. R. Gomes (✉)
School of Education, Polytechnic Institute of Coimbra, Coimbra, Portugal
e-mail: drmgomes@ipc.pt

Research Centre for Natural Resources Environment and Society (CERNAS), Coimbra, Portugal

A. Leandro
School of Education, Polytechnic Institute of Coimbra, NICSH—Research Group in Social and Human Sciences University of Minho—CECS—Communication and Society Research Center, Coimbra, Portugal

M. J. Santos
Lisbon School of Economics and Management—ISEG/Ulisboa, SOCIUS—Centro de Investigação em Sociologia e das Organizações, Lisbon, Portugal

C. Machado (ed.), *Building the Future with Human Resource Management*,
Management and Industrial Engineering, https://doi.org/10.1007/978-3-031-52811-8_3

environmental performance of the organisation. Furthermore, organisations applying GHRM practices more easily establish affective bonds with their employees.

Keywords Sustainability · Green management of human resources · Green behaviour · Green performance · Affective commitment

1 Introduction

Sustainability is a benchmark reference theme for the twenty-first century. The world is undergoing constant change: resources are scarce and the constant acts of man in the world renders the surrounding ecosystem increasingly fragile (driving issues such as global warming, pollution, among others). As such, effective and efficient actions are necessary to reversing this situation. Businesses and companies are therefore increasingly adopting sustainable practices across their different departments. This includes the Human Resource Department given its responsibility for deploying people across the organisation. Thus, the emergence of the green/environmental management of human resources (Green Human Resource Management—GHRM hereafter).

GHRM provides a concept interrelated with green (sustainable) practices subsequently implemented through Human Resource Management. The GHRM practices divide into the following categories: green recruitment and selection (e.g., applying environmental criteria in the recruitment and selection of persons displaying environmental awareness), green training and development (e.g., conveying information/knowledge about environmental themes and correspondingly developing competences in these areas), green performance management and evaluation (e.g., evaluating the performance of staff according to green criteria) and green rewards (e.g., providing financial and non-financial incentives to staff as means of recognition for the tasks/actions carried out whether directly or indirectly related with their positions/functions) (Opatha & Arulrajah, 2014).

Nowadays, people are more aware about the environmental questions and about how their acts may be prejudicial to those surrounding them (Geetha & Sammanasu, 2020). In the business context, the behaviour of each person has a direct/indirect impact on their organisation's development. Whenever a company adopts green practices (for example: a culture oriented towards environmental objectives and training), these seek to influence the behaviours of their staff and render them equally "eco-friendly" (Pham et al., 2019). Each behaviour manages an action that has a direct/indirect impact on an organisation's performance. This is susceptible to measurement through the positive/negative difference in the results obtained to those forecast (Elena-Iuliana & Maria, 2016). Clearly, there is every importance in organisations adopting increasingly active roles in their own environmental performances given that there is generally greater awareness/knowledge about the environmental impacts we each produce on the surrounding environment (Zeng et al., 2020). These same authors affirm there has been a change in the prevailing paradigm with investors opting to invest more in companies with positive environmental policies than those displaying a lack of concern over their environmental impacts on the surrounding environment. All these actions and behaviours generate consequences for the images of companies. This image is key to success and generating trust among staff and feelings of organisational pride and belonging (Figurska & Matuska, 2013). This feeling interrelates with the affective connections that staff maintain with their organisations, the affective commitment (Khaskheli et al., 2020).

There has been an abundance of research carried out by various researchers on this theme over the course of recent years. However, we found no research referencing GHRM practices and their impacts on the environmental behaviours and performance of both employees and the organisation in the Portuguese context. Hence, we encountered the first shortcoming in the literature, the lack of studies on the Portuguese case. Another failing of research stems from the overall lack of sectors explored within the scope of this theme (Bazrkar & Moshiripour, 2021; Zhu et al., 2021).

Hence, this study has the objective of understanding the impact of GHRM practices on the green behaviours of members of staff (eco-friendly behaviours), the environmental performance of organisations, and the affective commitment of their staff.

Literature review and formulation of hypotheses.

1.1 Green Human Resource Management

Green Human Resource Management (GHRM) incorporates all the set of activities undertaken under the auspices of human resource management with the objective of improving the environmental performance of organisations and encouraging employees to adopt green behaviours on an active and recurrent basis (Renwick et al., 2013). The same authors suggest that these GHRM practices begin when staff join the organisation and extend through to their departure.

As regards organisations, GHRM practices need integrating into the long term strategy (Saeed et al., 2018). These practices divide into three phases that share the objective of improving the sustainable performance of organisations (Mousa & Othman, 2020). According to these authors, the first phase consists of attracting and selecting candidates who display interest in environmental issues before the second phase involves retaining, developing, and improving the competences/knowledge of staff. Finally, the third phase spans the definition of green remuneration and an evaluation system able to raise the awareness and commitment of employees towards environmental matters.

1.1.1 Green Recruitment and Selection

Green Recruitment and Selection (GRS) incorporates the process of recruiting and selecting candidates who display a commitment to environmental issues (Jabbour et al., 2008).

Throughout selection processes, especially during the interview phase, employers may verify the level of candidates in terms of their knowledge, awareness, concern, and commitment towards the environment through raising and discussing such issues (Saeed et al., 2018).

GRS spans three core facets: green awareness, green employer branding, and green criteria (Jabbour, 2013; Renwick et al., 2013). Green awareness interrelates with the personality characteristics of each candidate that best enable the organisation to achieve its environmental objectives (Saeed et al., 2018). Staff with high levels of green awareness tend to get more actively involved in the environmental dimensions of companies as well as acting to deepen their environmental knowledge that correspondingly assists their companies turn in better environmental performances (Jehan et al., 2020; Perron et al., 2006). Green employer branding encapsulates the image and reputation of an organisation in relation to environmental issues as verified through their actions, especially their GHRM practices (Jackson et al., 2011). Employees at organisations with good brand and green reputations tend to feel pride in their place of work and being part of the project (Jackson et al., 2011; Saeed et al., 2018). Therefore, organisations attaining good green brands may more effectively attract and select potential candidates with positive mentalities towards environmental issues (Saeed et al., 2018). Green criteria serve to select from the different candidates (Renwick et al., 2013). In this aspect, organisations may include environmental facets in the descriptions of functions, raising questions related to their environmental knowledge, values, and beliefs in order to select better by incorporating these requirements (Renwick et al., 2013).

1.1.2 Green Training and Development

Green Training and Development (GTD) spans a set of different training and learning actions and activities that motivate staff to boost their skills, competences, and knowledge in relation to the environment and that result in greater awareness and concern among employees about environment related issues (Jabbour, 2011). Hence, organisations need to play a crucial role in fostering the awareness of staff, guiding them to think and deal with their work in more sustainable approaches (Zhu et al., 2021). Therefore, such training sets out with the objective of altering the attitudes and emotionally involving members of staff in environmental issues (Fernández et al., 2003).

GTD needs to extend to every employee in the organisation and not only the members of environment departments (Jehan et al., 2020; Saeed et al., 2018). To this end, the human resource department should carry out analysis to ascertain the learning components that need developing and/or learning (Saeed et al., 2018). In summary,

GTD holds importance to organisations in keeping with its capacity to stimulate and promote environment-friendly attitudes and behaviours among employees (Jehan et al., 2020; Saeed et al., 2018).

1.1.3 Green Performance Management and Evaluation

Green Performance Management and Evaluation (GPME) consist of an employee performance evaluation system spanning the sustainable processes and activities of the respective organisation (Jabbour et al., 2008). To this end, green performance indicators establish green criteria for all members of staff subject to evaluation with issues ranging from environmental responsibilities, communicating concerns about environmental policies, among others (Saeed et al., 2018). One of the core GPME objectives is to ensure that the results align with the sustainability objectives of the organisation (Mousa & Othman, 2020). Should the results diverge from those intended, organisations then need to verify just which aspects failed to go as planned and correspondingly review the originally defined sustainability plan (Mousa & Othman, 2020). An important factor for achieving the planned objectives and ensuring positive GPME results is the provision of regular feedback to employees about the ongoing results (Saeed et al., 2018).

1.1.4 Green Incentives and Remunerations Systems

The green incentive and remunerations system (GIRS) consists of a monetary and non-monetary remuneration system designed to attract, retain, and motivate staff to contribute towards achieving the environmental objectives set by the organisation (Jabbour et al., 2013; Jackson et al., 2011). When staff receive recompense for their environmental performance within an organisation, they motivate other members to follow the same path and engage in eco-friendly behaviours (Saeed et al., 2018). Jabbour et al. (2008) affirm that staff feel more motivated when receiving non-monetary benefits as is the case of recognition and praise.

1.2 Individual Staff Green Behaviours

Employees' pro-environmental behaviours defines the autonomous behaviours of an organisation that contribute to building sustainable development into the workplace (Ones & Dilchert, 2012). Some of these behaviours were identified by Saeed et al. (2018): turning the office lights off when not in use, switching from commuting by car to bicycle, reducing excesses and waste, among others.

Green behaviours may be considered as pro-environmental behaviours and divide into two broad types: (1) those related to the task in which the type of behaviour interrelates with daily tasks of a sustainable nature and as demanded by the organisation, for example, compliance with environmental responsibilities and norms; and (2) voluntary green behaviours, thus, autonomous and voluntary actions of an environmental character, for example, printing on both sides of paper or encouraging colleagues to act on environmental questions (Zhu et al., 2021). "The success of organisational initiatives for environmental sustainability depend on the environmental behaviours of their staff" (Saeed et al., 2018, p.424). In turn, green behaviours may be divided into three dimensions: green organisational citizenship behaviours, green interpersonal citizenship behaviours, and official green behaviours (Opatha & Arulrajah, 2014). Green organisational citizenship behaviours reflect actions/behaviours that do not fall within the functions of the employee and that the individual undertakes autonomously and voluntarily. Examples of these behaviours are the following: travelling to work via less polluting means (on foot, by bike, or shared car), reusing paper, turning off equipment when not in use, among others; green interpersonal citizenship behaviours span actions /behaviours that employees engage in that influence their peers to also undertake green behaviours. These are engaged in voluntarily and do not fit directly into their respective functions. One example of such behaviours would be teaching peers to adopt green behaviours and explain the positive effects of their actions. In turn, official green behaviours convey the behaviours/actions that members of staff should necessarily carry out as they fall within the description of their functions; (Opatha & Arulrajah, 2014). In summary, employees displaying green behaviours avoid any alterations that might harm the surrounding environment, display care in the utilisation of natural resources, avoid/minimise polluting the environment through their actions and display behaviours that make their surrounding environments healthier (Opatha & Arulrajah, 2014).

GHRM practices enable members of staff to adopt environment-friendly behaviours in both their personal and their professional lives (Ansari et al., 2020; Ribeiro et al., 2022; Saeed et al., 2018). Furthermore, out of the many advantages generated by GHRM practices, the reduction in costs and improvement to the organisation-employee relationship particularly stand out (Ansari et al., 2020). In the business context, GHRM practices influence the behaviour of staff for various reasons, especially: (1) they are immediately introduced to the sustainable/environmental ideologies of the company right from their first contact, thus, during the recruitment and selection process (in practice, conveying to new recruits the importance the company attributes to the environment and its own particular sustainable values); (2) training that enables staff members to develop their own competences, knowledge and attitudes in relation to environmental issues; (3) remuneration reflects and recognises the positive performance of employees that then motivate them to continue improving their performance as well as participating in the ongoing sustainable activities in the company (Renwick et al., 2013).

As regards green recruitment, this highlights the commitment of organisations to act in the interest of the environment, attracting and retaining individuals with the same convictions (Ojo et al., 2020). During the recruitment and selection process, employees are from the outset made aware of the commitments made by the company in terms of sustainability and correspondingly adopt behaviours that align with them (Renwick et al., 2013).

Another GHRM practice involves training processes that raise the awareness of employees as regards the importance and the means of environmental protection, deepening their understanding of environmental prevention issues (Saeed et al., 2018). Staff need to have the opportunity to participate in green initiatives as well as the scope for staff to propose solutions to the environmental problems the company faces (Chaudhary, 2019). As regards performance management, when staff do not attain the green criteria evaluated by the organisation (for example: reduction of emissions), the organisation should encourage staff to attain them in the future, thus by adopting green behaviours (Tang et al., 2017). Another important practice influencing the behaviour of staff is green remuneration. When staff receive financial or non-financial rewards for good environmental performance, this also motivates the staff around them to adopt green behaviours (Saeed et al., 2018). Furthermore, when such behaviours "are officially appreciated and recognised, this then becomes common practice in the workplace" (Chaudhary, 2019, p.633). Indeed, this happens as members of staff, on gaining recognition, feel pride in their actions and this nurtures their wellbeing and self-esteem and thereby continuing to repeat the rewarded action/task (Jehan et al., 2020). A study carried out by Bazrkar and Moshiripour (2021) concludes that GHRM practices are a factor of success in transforming regular employees into employees displaying green behaviours. Whenever an organisation maintains working environments focused on GHRM practices, staff tend to ensure their behaviours align with these same practices (Ansari et al., 2020).

According to the literature, various studies identify the positive effects of GHRM practices on green behaviour in the international context (Ansari et al., 2020; Bazrkar & Moshiripour, 2021; Chaudhary, 2019; Jehan et al., 2020; Ojo et al., 2020). Hence, we arrive at our first study hypothesis:

H1: GHRM practices have a positive impact on employees' green behaviour

1.3 Company Green Performance

The performance of a company may be measured through their results (through recourse to the balance sheet, accountancy reporting, key performance indicators, among others), which may be either positive or negative (Elena-Iuliana & Maria, 2016).

Wagner et al. (2002) refer to how the environmental performance of an organisation conveys its behaviour in relation to the surrounding environment as a means of controlling its actions in terms of the consumption of the resources it has available alongside concerns over pollution emissions, levels of waste, among other factors (Singh et al., 2019, p. 205). Currently, in addition to economic responsibility, the sustainable and social responsibilities of companies are increasingly prominent (Jehan et al., 2020).

The behaviour of staff is essential to achieving the desired levels of environmental performance as well as to subsequent improvements (Úbeda-García et al., 2021). To this end, senior management need to pay attention to the level of ecological knowledge that their employees hold in order to be able to achieve the ideal level of performance. Whether in the format of continuous training, in selection and recruitment and as well as in the "definition of the performance evaluation criteria of staff based on their level of knowledge and awareness of environmental programs provide organisations with the conditions appropriate to the continuous improvement of their environmental performance" (Bazrkar & Moshiripour, 2021, p. 103).

As already referenced, there are GHRM practices that enable improvements to the environmental performance of organisations, including green selection, green training, and development, green evaluation and green reward systems (Mousa & Othman, 2020). Initially, whenever organisations integrate a focus on good environmental performance into their strategy, they pay particular attention during the phase of recruitment and selection, ensuring candidates display environmental awareness and a commitment to improving the environmental performance of the organisation (Jackson et al., 2011). When companies identify functions/tasks for staff to perform with the objective of improving their environmental performance and provide monetary/non-monetary rewards for obtaining the set objectives, then the environmental performance improves (Jehan et al., 2020). As regards training (focused on those environmental aspects the company needs to improve), when the knowledge/competences/attitudes acquired are put into practice, this helps improve the respective company's environmental performance (Jehan et al., 2020). Green performance evaluations enable constant feedback that supports the continuous improvement of company environmental performance standards (Jackson et al., 2011). Some examples of green criteria applied in performance evaluations, which directly influence the environmental performance of organisations, include: "environmental incidents, environmental responsibilities, reduction in carbon emissions and communicating environmental concerns and policies" (Tang et al., 2017, p.36).

In accordance with the existing literature, various studies identify the positive effects arising from GHRM practices on the green performance standards of organisations (Jackson et al., 2011; Jehan et al., 2020; Mousa & Othman, 2020). Hence, we may put forward our second study hypothesis:

H2: GHRM practices have a positive impact on company green performance

1.4 Affective Commitment

Commitment may be depicted through the relationships employees maintain with their organisations (Meyer & Allen, 1991). The definition of this relationship incorporates three distinctive feelings: desire (affective commitment), need (commitment to continuity) and obligation (normative commitment) (Meyer & Allen, 1991). The commitment to continuity reflects the costs associated with employees terminating their status, thus, employees become dependent on continuity in the organisation due to important aspects (salary, pension, length of service, stability, etcetera) (Meyer & Allen, 1991). Normative commitment arises when employees feel a duty/obligation to remain in their organisations (Meyer & Allen, 1991).

Some scholars reveal that organizational support, authentic leadership and psychological capital are predictors of affective commitment (e.g., Ribeiro, et al., 2020; Semedo et al., 2018; Yücel et al., 2020).

Affective commitment encapsulates the emotional connection that employees hold towards their organisation (Khaskheli et al., 2020). This concept approaches the emotions, such as affection and belonging, that employees feel towards their organisation (Bouraoui et al., 2018). According to these authors, members of staff that display affective commitment register a greater probability of remaining longer in an organisation of their own free and spontaneous will. Furthermore, when an employee has a strong connection with an organisation, they feel pride over their membership (Khaskheli et al., 2020).

Furthermore, affective commitment may influence the intentions of employees as regards changing jobs as the higher the level of affective commitment of staff towards their organisations, the greater the commitment and the lower the level of staff turnover (Gyensare et al., 2016).

Hence, members of staff who hold environmental values attuned to those of the organisation display a greater propensity to be more affective (Ashforth & Mael, 1989). This arises because they identify with and feel integrated into a company with a good reputation and image as regards the surrounding environment (Tajfel & Turner, 1985). Organisations that engage in GHRM practices report that implementation fosters more positive self-images (Farooq et al., 2014).

In keeping with the existing literature, this seeks to verify the impact of GHRM practices on affective commitment in accordance with social identity theory (Mael & Ashforth, 1995). We therefore arrive at our third hypothesis:

H3: GHRM practices generate a positive impact on the affective commitment of employees.

Conceptual research model.

Taking into account the hypotheses set out above, the objective of this study involves understanding the relationship between GHRM practices, Green Behaviour, Green Performance, and Affective Commitment. The expectation is to return a positive relationship between GHRM practices and Green Behaviour, Green Performance and Affective Commitment. Thus, we may set out the following research model (Fig.1).

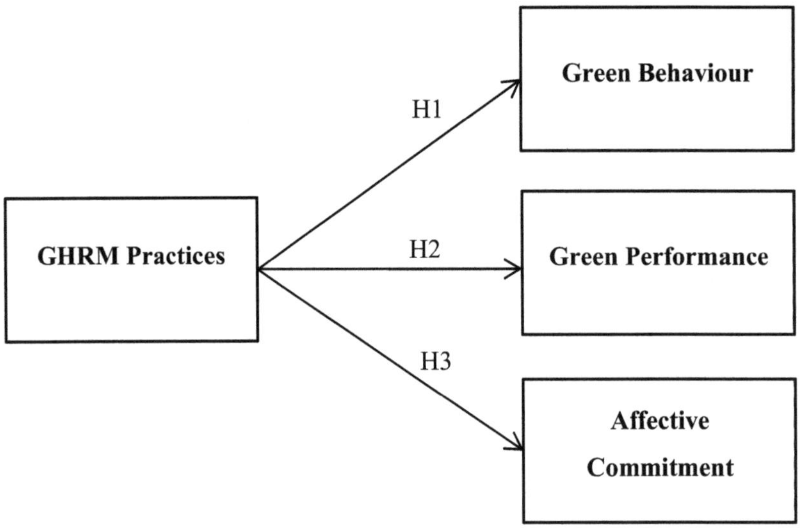

2 Method

2.1 Data Collection Procedure and Constructing the Sample

The present study adopts a cross-sectional approach given the data gathered details the particular time of its collection. Furthermore, this study is a quantitative study with the objective of verifying the relationship between GHRM practices and the green behaviours, green performance and affective commitment of employees.

We applied a primary data source having made recourse to an online questionnaire distributed via Google Docs platform and subsequently highlighted via email and the LinkedIn social network. The questionnaire featured scales already validated by other authors with the option of applying an online questionnaire inspired by the already existing literature (Bazrkar & Moshiripour, 2021; Jehan et al., 2020; Saeed et al., 2018; Zhu et al., 2021).

The gathering of data took place at the beginning of 2022 (over three months) and reached out to managers/staff in human resource departments given this study applies to the practices implemented by this department. We would here highlight that the questionnaire was completed anonymously and with all the information entirely confidential and applied exclusively to the present study.

In order to process and analyse the data, we deployed the IBM SPSS Statistic software.

2.2 Questionnaire Structure and Measurement Instruments

The questionnaire applied for this study divides into six sections. The first features an introduction which presents the scope and purpose of the questionnaire as well as guaranteeing the confidentiality of data and requesting their due authorised consent. The remaining sections deal with each of the variables subject to measurement before the final section collects socio-demographic data.

In order to analyse and test the veracity of the hypotheses, we applied measurement scales already validated in the existing literature and with a 5-point Likert scale as the means of responding. We would highlight the application of two Likert scales (cf. Annexes, Table 7). The second section sets out statements in relation to the "GHRM practices" variable, measured through a scale adapted and translated from the study by Kim et al. (2019), made up of six items. To respond to these items, we applied a 5-point Likert scale version 1 (cf. Annexes, Table 8). The third section contains seven statements against which participants evaluated their "Green Behaviour" measured through a scale adapted and translated from the study by Kim et al. (2019), including a total of seven items. To respond to these items, we applied a 5-point Likert scale version 2 (cf. Annexes, Table 9). The fourth section relates to the "Green Performance" of the organisation variable. In this section, the participant responds to seven statements relating to specific actions undertaken by their organisations over the last six months, measured through a scale adapted and translated from the study by Kim et al. (2019) containing a total of seven items. To this end, we applied the 5-point Likert scale version 2 (cf. Annexes, Table 10). The fifth section applies three statements about the "Affective Commitment" variable. This serves to evaluate the feelings of the participant as regards the situations presented based on a scale adapted and translated from the study by Rego et al. (2011), formed by 3 items. In order to respond, participants answered according to the 5-point Likert scale version 2 (cf. Annexes, Table 11).

2.3 Sample Characteristics

The sample for this study contains 212 individual members of Human Resource Departments, of whom 142 (67%) are female and with 70 (33%) male respondents (cf. Annexes, Table 12), with an average age of 43 (cf. Annexes, Table 13). As regards their educational qualifications, three-quarters of the sample (75.9%) reported higher education (undergraduate or higher degree) (cf. Annexes, Table 14). In keeping with the national business sector, the sample contains only 37 (17.5%) participants working in large companies with the remainder employed at micro, small and medium sized companies (cf Annexes, Table 15). The companies in the sample primarily have their headquarters in the districts of Lisbon (15.6% of the sample), Braga (10.4%), Oporto (10.4%), Leiria (9.4%), and Aveiro (8.5%) (cf. Annexes, Table 16).

Table 1 Internal consistency of variables

	Cronbach's Alfa	Number of items
GHRM practices	0.921	6
Green Behaviour	0.740	7
Green Performance	0.919	7
Affective Commitment	0.947	3

Source Authors

2.4 Internal Consistency of the Scales

We measured the internal consistency of the scales through Cronbach's Alpha. This served to measure the reliability of the items making up the scales for the variables under study. We would note that Cronbach's Alpha results may vary between 0 and 1, with 0.6 representing the acceptable minimum (Maroco & Garcia-Marques, 2006).

Analysing Table 1, we may verify that the variables GHRM practices, Green Performance and Affective Commitment return Cronbach's Alfa results of over 0.9, which reflects a very good level of internal consistency. We would furthermore note that while the Green Behaviour variable registers a level below 0.9, this variable nevertheless returns a good Alfa result (good internal consistency).

3 Results: Discussion and Analysis

The present study took place with the objective of responding to the research question formulated above: what are the impacts of GHRM practices on the perceptions, attitudes, and behaviours of employees, particularly green behaviours, the perceptions of employees about the green performance of their organisation and as regards their own affective commitment. This section analyses the descriptive statistics returned for the variables, their correlations, and linear regressions. Finally, we put forward a reflection on the results obtained.

3.1 Descriptive Analysis and Correlations

To analyse the variables, as detailed above, we applied a 5-point Likert scale.

Analysing Table 2 verifies how GHRM practices are the variable returning the lowest average (2.5849), which indicates how few companies actually implement this type of practice. In terms of the respondents, one-half of the sample answered below or equal to 2.5 (3) and with the other half of the sample providing answers equal to or higher than 3. As regards green behaviour, this returns an average of 4.4643 (on a scale of 1 to 5), which indicates how respondents hold high levels

Table 2 Descriptive statistics of the variables studied

	Average	Median	Standard Deviation	Minimum	Maximum
GHRM practices	2.5849	2.5	1.12788	1	5
Green Behaviour	4.4643	4.5714	0.52057	2.29	5
Green Performance	3.4784	3.5714	0.87998	1	5
Affective Commitment	4.1714	4.1667	0.92276	1	5

Response scale from 1 to 5 with 1 the weakest level of agreement and 5 the strongest
Source Authors

of environmental awareness and, correspondingly, display green behaviours in the workplace. Green performance returns an average of 3.4784, which reflects how the companies surveyed return good levels of environmental performance even if this could be better. Furthermore, this average broadly reflects the general prevalence of GHRM practices, which suggests that even though companies do now widely undertake GHRM practices, their environmental performance is also not the very best (in the perception of the respective member of staff). Finally, affective commitment returns an average of 4.1714, which seems to demonstrate how respondents experience a high level of pride at belonging to their organisations.

The GHRM practices variable reports a standard deviation rate of (sd) = 1.12788, Green Behaviour came in at sd = 0.52057, Green Performance sd = 0.87998, and with Affective Commitment on sd = 0.92276. The standard deviations return lower values than the average which indicates a low level of data dispersion.

Analysing the impact of GHRM practices on Green Behaviour verifies that they return a Spearman coefficient result of (r) = 0.313** and $p < 0.01$, which suggests that both variables are positively related, thus, when one rises/falls, the other also rises/falls in the same direction. In practice, this means that the greater the extent of GHRM practices implemented in organisations, the greater the adoption of green behaviours by their respective members of staff. The same also holds in the opposite direction: the lesser the implementation of green practices, the lower the level of green behaviours.

As regards the impact of GHRM practices on environmental performance, the results indicate that these two variables return results of r = 0.564** and $p < 0.01$, which reflects how both positively interrelate. The greater the GHRM practices, the higher the organisational environmental performance and, in turn, the lower the level of GHRM practices, the lower the impact on the environmental performance.

Finally, as regards the impact of GHRM practices on affective commitment, the results, r = 0.326** and $p < 0.01$, demonstrate that both variables positively interact and therefore, the greater the extent of GHRM practices in a company, the greater the affective performance of its staff and vice versa with lower levels GHRM practices driving lower affective performance results.

In summary, and after analysing Table 3, we may state that the variables display a strong and mutual correlation.

Table 3 Spearman's coefficient

	1	2	3	4	5	6	7
1-Gender (a)	1						
2-Academic qualifications (b)	0.049	1					
3-Company size (c)	0.021	-0.024	1				
4-GHRM practices	0.048	0.116	-0.156*	1			
5-Green Behaviour	-0.147*	0.116	-0.083	0.313**	1		
6-Green Performance	-0.049	0.055	-0.178**	0.564**	0.380**	1	
7-Affective Commitment	-0.042	0.054	-0.234**	0.326**	0.136**	0.247**	1

*This correlation is significant at 0.05 (2 decimal places)
** This correlation is significant at 0.01 (2 decimal places)
(a): 1- Female; 2- Male
(b): 1- Primary School; 2- Secondary School; 3- Undergraduate Degree; 4- Master's Degree; 5- Doctoral Degree
(c): 1-Micro Company; 2-Small Company; 3-Medium Company; 4-Large Company
Source Authors

In relation to the socio-demographic variables, the results demonstrate that only two return significant levels of correlation. Company size correlates negatively and significantly with Green Performance ($r = -0.178^{**}$ and $p < 0.01$) and Affective Commitment ($r = -0.234^{**}$ and $p < 0.01$) variables that suggests larger companies tend to register lower levels of environmental and affective commitment.

3.2 *Relationships Existing Among the Variables*

In order to verify the impact of GHRM practices on the green behaviour, environmental performance and affective performance, we deployed linear regressions to evaluate the underlying assumptions (Pestana & Gageiro, 2014).

3.2.1 Relationship Between GHRM Practices and Green Behaviour

As regards the first hypothesis stating that GHRM practices generate positive impacts on the green behaviours of employees, we carried out a linear regression.

As Table 4 sets out: 9% of the total variability in green behaviour stems from GHRM practices. Furthermore, the results verify that for each additional unit of GHRM practices, Green Behaviour rises by an average of 0.142. In addition, the results display an F value of 21.901 with a level of significance of less than 0.001.

As the p value result is very low (bordering on 0), this means that there is statistical evidence that the relationship between these two variables is significant and positive (Beta = 0.142).

Hence, the results provide support to H1.

Table 4 Linear regression results (H1)

	Non-standard Beta	F Test	Adjusted R^2	Level of significance
GHRM practices—Green Behaviour	0.142	21.901	0.090	< 0.001

Source Authors

Table 5 Linear regression results (H2)

	Non-standard Beta	F Test	Adjusted R^2	Level of significance
GHRM practices—Green Performance	0.434	93.963	0.306	< 0.001

Source Authors

3.2.2 Relationship Between GHRM Practices and Green Performance

As regards the second hypothesis ascertaining whether GHRM practices return a positive impact on the Green Performance of employees, we undertook a linear regression.

Analysing Table 5, we may verify the following: 30.6% of the variation in Green Performance derives from GHRM practices. The results furthermore find that for each additional unit of GHRM practices, the Green Performance level rises by an average of 0.434. This also produced an F value of 93.963 with a level of significance of below 0.001.

The very low level of the p value (close to 0) means that there is statistical evidence that the relationship between the two variables is both significant and positive (Beta = 0.434).

Thus, the results lend support to H2.

3.2.3 Relationship Between GHRM Practices and Affective Commitment

For the third hypothesis that posits GHRM practices generating positive impacts on the Affective Commitment of staff, we carried out a linear regression.

Analysis of Table 6 reports the following: 9.6% of the variability in Affective Commitment stems from the prevailing GHRM practices. This furthermore confirms that each additional unit of GHRM practices raises the Affective Commitment by an average of 0.259. The regression also returned an F value of 23.354 with a level of significance of below 0.001.

The very low level of p value (close to 0) conveys the presence of statistical evidence that the relationship between these two variables is significant and positive (Beta = 0.259).

Table 6 Linear regression results (H3)

	Non-standard Beta	F Test	Adjusted R^2	Level of significance
GHRM practices—Affective Commitment	0.259	23.354	0.096	< 0.001

Source Authors

Hence, the study results lend support to H3.

3.3 Discussion

As regards the hypotheses here subject to analysis, the results obtained report that: (1) as regards hypothesis 1, which evaluates the impact of GHRM practices on the individual green behaviours of members of staff, receives support given how the variable strongly correlates and attains high levels of significance. Hence, we may state that this result aligns with some of the studies existing in the literature (Ansari et al., 2020; Bazrkar & Moshiripour, 2021; Chaudhary, 2019; Jehan et al., 2020; Ojo et al., 2020). When an organisation deploys GHRM practices that are recognised and identified by staff members, the latter tend to adopt green behaviours in keeping with such practices (Ansari et al., 2020); (2) in terms of hypothesis 2, approaching the impact of GHRM practices on the environmental performance of organisations, we may verify that both variables strongly correlate and obtain significance. Furthermore, this also reports that the Green Performance variable is that which displays the greatest level of variation subject to explanation by GHRM practices (R2 adjusted = 30.6%). Consequently, this hypothesis receives verification and corroboration and also in keeping with the existing literature (Jackson et al., 2011; Jehan et al., 2020; Mousa & Othman, 2020); (3) in terms of hypothesis 3, which evaluates the impact of GHRM practices on the affective commitment of employees, this also receives support as the variables display a strong mutual correlation in a significant relationship. This result aligns with the literature review, especially as regards social identity theory (Mael & Ashforth, 1995) that explains the relationship between organisations and their members of staff (e.g., when an organisation displays positive values, as is the case with environmental concerns, employees tend to report relationships of affective commitment towards their organisation). Hence, companies taking on positive values in which employees feel pride and identify with ensures the latter more easily engage in their respective company's practices, as is the case with GHRM practices, and thereby foster and improve their affective commitment within the scope of their organisations.

In summary, GHRM practices generate positive impacts on the variables under study and this reflects how the greater the extent of GHRM practices prevailing in any organisation, the stronger the green behaviours and affective commitment of staff in conjunction with improvements to the environmental performance of organisations.

4 Conclusion

4.1 Key Conclusions and Contributions to Management

The present study enabled the response to the objective initially set out: verifying whether GHRM practices produced positive impacts on the green behaviours and the affective commitment of employees as well as the green performance of their organisation. After analysing the key results obtained, we may conclude that when organisations deploy effective GHRM practices, they positively shape and influence the aforementioned variables.

As regards the GHRM practices and green behaviour, the conclusions portray how when organisations apply GHRM practices in their day-to-day operations, members of staff tend to follow these same practices and therefore adopt (whether consciously or subconsciously) green behaviours. Furthermore, the same applies in the contrary direction and, therefore, when companies lack any GHRM practices, staff do not engage in ecological behaviours. When implemented within an organisation, these practices enable diverse advantages to the organisation, especially in terms of reducing cost and wastage (Ansari et al., 2020; Saeed et al., 2018). Hence, we may state that these practices foster positive behaviours from employees, which in turn shapes the vision employees hold about their organisation as well as its overall image and reputation.

As regards GHRM practices and green performance, this study concludes that the greater the extent of GHRM practices in an organisation, the better its environmental performance. The same also prevails in the opposite sense (the fewer the GHRM practices, the lower the environmental performance). Currently, the question of sustainability is a theme present in the reality of each company (Gomes et al., 2023) given the increasing scarcity of resources and the rising pressures for companies to reduce their impacts on the surrounding environment. Hence, the environmental impact of any organisation may begin to be reduced through the actions of its members, thus the employees. When staff are aware of the impact of their actions on the environment, they tend to reduce their harmful actions. Some of the means of building up the environmental performance through GHRM practices include the following: recruiting staff who display a positive awareness as regards sustainability, rewarding staff who effectively implement GHRM practices, engaging in the evaluation of green performance adjusted by the green behaviours of employees, among others.

Finally, as regards GHRM practices and affective commitment, we may conclude that the greater the extent of GHRM practices, the higher the level of employee affective commitment. As already mentioned, issues around sustainability are crucial issues to the daily lives of each person. Nowadays, people are more aware, also due to the campaigning efforts carried by the media in different formats, about the impacts of actions on the environment. As such, staff belonging to organisations that display clear environmental concerns feel pride in their connection with the organisation and this raises their desire to remain in their positions.

Table 7 Scales applied

Study variables	Scale applied
GHRM practices	5-point Likert scale version 1
Green Behaviour	5-point Likert scale version 2
Green Performance	5-point Likert scale version 2
Affective Performance	5-point Likert scale version 2

Source Authors

4.2 Limitations and Suggestions for Future Research

The current study displays certain limitations that are themselves appropriate for consideration by future research. Despite the sample containing a considerable number of participants, the data collection time was greater than that expected due to lengthy response times because of the pandemic and as a proportion of the data collection period overlapped with the beginning of the year. We would note here that the sample was obtained by convenience and the data findings may thus not be generalised.

Furthermore, another limitation derives from the limited scope of studies that relates GHRM with Affective Commitment variable which, while hindering comparisons, reflects an innovative contribution made by this study. Another limitation arises from the scope for contamination by variance error in the common methodology given we collected data from the same source and in a singular defined timeframe (Podsakoff et al., 2003).

Despite these limitations, we fully consider that the results obtained contribute to enriching both the literature and organisations.

One avenue for future research, in addition to including other variables, involves studying the impact of GHRM practices on employer branding. The literature already contains some studies interrelating these two variables and it would be thus of relevance to verify this relationship within the Portuguese context and thereby establish comparisons with the international context.

5 Annexes

See Tables 7, 8, 9, 10, 11, 12, 13, 14, 15 and 16.

Table 8 Items in the GHRM practices variable

1.	My company provides appropriate training for promoting green/ environmental management as a fundamental organisational value
2.	My company considers "environment-friendly" employees as part of their performance evaluation
3.	My company associates the ecological behaviours of employees to their rewards and remuneration
4.	My company considers the balance between personal identity and environmental management in processes of recruitment and selection
5.	My company undertakes initiatives so that employees perfectly understand the extent of the company's environmental policy
6.	My company promotes and encourages employees to provide suggestions on how to improve the workplace environment

Source Authors

Table 9 Items in the Green Behaviour variable

1.	Before leaving my workplace, I turn off all electrical devices, such as computers, television monitors, etc
2.	In my workplace, when I leave an empty location, I turn off the light
3.	I separate and recycle waste and rubbish in my workplace
4.	I use materials conservatively in my workplace
5.	I reuse materials in my workplace
6.	I limit my usage of water in the bathroom in order to save this resource
7.	I pay significant attention to the wastage of water

Source Authors

Table 10 Items in the Green Performance variable

1.	There was a reduction in waste
2.	There was conservation in water consumption
3.	There was conservation in energy consumption
4.	There was a reduction in the purchase of materials, chemical products and non-renewable components
5.	There was a reduction in general costs
6.	There was an improvement in market positioning
7.	This helped improve the company's reputation

Source Authors

Table 11 Items in the Affective Performance variable

1.	I am proud to tell other people I'm a member of this company
2.	I have a strong emotional bond with this organisation
3.	In this company, I feel like "part of the family"

Source Authors

Table 12 Sample distribution by gender

	Frequency	Percentage (%)
Female	142	67
Male	70	33
Total	212	100

Source Authors

Table 13 Descriptive statistic of participant ages

Variable	Age
Average	42.89
Mode	44
Median	44
Youngest	21
Eldest	65

Source Authors

Table 14 Sample distribution by academic qualification

	Frequency	Percentage (%)
Primary school	0	0
Secondary school	36	17
Undergraduate degree	125	58.9
Master's degree	50	23.6
Doctoral degree	1	0.5
Total	212	100

Source Authors

Table 15 Sample distribution by size of participating company

	Frequency	Percentage (%)
Micro company	42	19.8
Small company	62	29.2
Medium company	71	33.5
Large company	37	17.5
Total	212	100

Source Authors

Table 16 Sample distribution by district

	Frequency	Percentage (%)
Aveiro	18	8.5
Beja	4	1.9
Braga	22	10.4
Castelo Branco	4	1.9
Coimbra	14	6.6
Évora	5	2.4
Faro	13	6
Guarda	14	6.6
Leiria	20	9.4
Lisboa	33	15.6
Portalegre	1	0.5
Porto	22	10.4
Santarém	13	6.1
Setúbal	11	5.2
Viana do Castelo	1	0.5
Vila Real	3	1.4
Viseu	9	4.2
Autonomous Region of the Azores	4	1.9
Autonomous Region of Madeira	1	0.5
Total	212	100

Source Authors

References

Ansari, N. Y., Farrukh, M., & Raza, A. (2020). Green human resource management and employees pro-environmental behaviours: Examining the underlying mechanism. *Corporate Social Responsibility and Environmental Management, 28*(1), 229–238. https://doi.org/10.1002/csr.2044

Ashforth, B. E., & Mael, F. (1989). Social identity theory and the organization. *Academy of Management Review, 14*(1), 20–39. https://doi.org/10.5465/amr.1989.4278999

Bazrkar, A., & Moshiripour, A. (2021). Corporate practices of green human resources management. *Foresight and STI Governance, 15*(1), 97–105. https://doi.org/10.17323/2500-2597.2021.1.97.105.

Bouraoui, K., Bensemmane, S., Ohana, M., & Russo, M. (2018). Corporate social responsibility and employees' affective commitment: A multiple mediation model. *Management Decision, 57*(1), 152–167. https://doi.org/10.1108/md-10-2017-1015

Chaudhary, R. (2019). Green human resource management and employee green behavior: An empirical analysis. *Corporate Social Responsibility and Environmental Management, 27*(2), 630–641. https://doi.org/10.1002/csr.1827

Elena-Iuliana, I., & Maria, C. (2016). Organizational performance—A concept that self-seeks to find itself. *Annals—Economy Series, 4,* 179–183.

Farooq, O., Payaud, M., Merunka, D., & Valette-Florence, P. (2014). The impact of corporate social responsibility on organizational commitment: Exploring multiple mediation mechanisms. *Journal of Business Ethics, 125*(4), 563–580. https://doi.org/10.1007/s10551-013-1928-3
Fernández, E., Junquera, B., & Ordiz, M. (2003). Organizational culture and human resources in the environmental issue: A review of the literature. *The International Journal of Human Resource Management, 14*(4), 634–656. https://doi.org/10.1080/0958519032000057628
Figurska, I., & Matuska, E. (2013). Employer branding as a human resources management strategy. *Human Resources Management & Ergonomics, 7*(2), 35–51.
Geetha, U., & Sammanasu, J. M. (2020). Green HRM-A conceptual framework. *Journal of xi'an University of Architecture & Technology, 7*(5), 1204–1212.
Gomes, D. R., Ribeiro, N., & Santos, M. J. (2023). "Searching for Gold" with sustainable human resources management and internal communication: Evaluating the mediating role of employer attractiveness for explaining turnover intention and performance. *Administrative Sciences, 13*, 24. https://doi.org/10.3390/admsci13010024
Gyensare, M. A., Anku-Tsede, O., Sanda, M. A., & Okpoti, C. A. (2016). Transformational leadership and employee turnover intention: The mediating role of affective commitment. *World Journal of Entrepreneurship, Management and Sustainable Development, 12*(3), 243–266. https://doi.org/10.1108/WJEMSD-02-2016-0008
Jabbour, C. J. C. (2011). How green are HRM practices, organizational culture, learning and teamwork? A Brazilian study. *Industrial and Commercial Training, 43*(2), 98–105. https://doi.org/10.1108/00197851111108926
Jabbour, C. J. C. (2013). Environmental training in organisations: From a literature review to a framework for future research. *Resources, Conservation and Recycling, 74*, 144–155. https://doi.org/10.1016/j.resconrec.2012.12.017
Jabbour, C. J. C., Santos, F. C. A., & Nagano, M. S. (2008). Environmental management system and human resource practices: Is there a link between them in four Brazilian companies? *Journal of Cleaner Production, 16*(17), 1922–1925. https://doi.org/10.1016/j.jclepro.2008.02.004
Jackson, S. E., Renwick, D. W. S., Jabbour, C. J. C., & Muller-Camen, M. (2011). State-of-the-art and future directions for green human resource management: Introduction to the special issue. *German Journal of Human Resource Management, 25*(2), 99–116. https://doi.org/10.1177/239700221102500203
Jehan, Y., Hussai, D., Batool, M., & Imran, M. (2020). Effect of green human resource management practices on environmental sustainability. *International Journal of Human Capital in Urban Management, 5*(2), 153–164. https://doi.org/10.22034/IJHCUM.2020.02.06.
Khaskheli, A., Jiang, Y., Raza, S. A., Qureshi, M. A., Khan, K. A., & Salam, J. (2020). Do CSR activities increase organizational citizenship behavior among employees? Mediating role of affective commitment and job satisfaction. *Corporate Social Responsibility and Environmental Management, 27*(6), 2941–2955. https://doi.org/10.1002/csr.2013
Kim, Y. J., Kim, W. G., Choi, H.-M., & Phetvaroon, K. (2019). The effect of green human resource management on hotel employees' eco-friendly behavior and environmental performance. *International Journal of Hospitality Management, 76*, 83–93. https://doi.org/10.1016/j.ijhm.2018.04.007
Maroco, J., & Garcia-Marques, T. (2006). Qual a fiabilidade do alfa de Cronbach? Questões antigas e soluções modernas. *Laboratório de psicologia, 4*(1), 65–90. https://doi.org/10.14417/lp.763.
Meyer, J. P., & Allen, N. J. (1991). A three-component conceptualization of organizational commitment. *Human Resource Management Review, 1*(1), 61–89. https://doi.org/10.1016/1053-4822(91)90011-Z
Mousa, S. K., & Othman, M. (2020). The impact of green human resource management practices on sustainable performance in healthcare organisations: A conceptual framework. *Journal of Cleaner Production, 243*, 118595. https://doi.org/10.1016/j.jclepro.2019.118595
Ojo, A. O., Tan, C.N.-L., & Alias, M. (2020). Linking green HRM practices to environmental performance through pro-environment behaviour in the information technology sector. *Social Responsibility Journal, 18*(1), 1–18. https://doi.org/10.1108/SRJ-12-2019-0403

Ones, D. S., & Dilchert, S. (2012). Environmental sustainability at work: A call to action. *Industrial and Organizational Psychology, 5*(4), 444–466.

Opatha, H. H. D. N. P., & Arulrajah, A. A. (2014). Green human resource management: Simplified general reflections. *International Business Research, 7*(8), 101–112. https://doi.org/10.5539/ibr.v7n8p101

Perron, G. M., Côté, R. P., & Duffy, J. F. (2006). Improving environmental awareness training in business. *Journal of Cleaner Production, 14*(6–7), 551–562. https://doi.org/10.1016/j.jclepro.2005.07.006

Pham, N. T., Tučková, Z., & Phan, Q. P. T. (2019). Greening human resource management and employee commitment towards the environment: An interaction model. *Journal of Business Economics and Management, 20*(3), 446–465. https://doi.org/10.3846/jbem.2019.9659

Podsakoff, P. M., MacKenzie, S. B., Lee, J. Y., & Podsakoff, N. P. (2003). Common method biases in behavioral research: A critical review of the literature and recommended remedies. *Journal of Applied Psychology, 88*(5), 879–903. https://doi.org/10.1037/0021-9010.88.5.879

Rego, A., Ribeiro, N., Cunha, M. P., & Jesuino, J. C. (2011). How happiness mediates the organizational virtuousness and affective commitment relationship. *Journal of Business Research, 64*(5), 524–532. https://doi.org/10.1016/j.jbusres.2010.04.009

Renwick, D. W. S., Redman, T., & Maguire, S. (2013). Green human resource management: A review and research Agenda. *International Journal of Management Reviews, 15*(1), 1–14.

Ribeiro, N., Gomes, D. R., Ortega, E., Gomes, G. P., & Semedo, A. S. (2022). The impact of green HRM on employees' eco-friendly behavior: The mediator role of organizational identification. *Sustainability, 14*, 2897. https://doi.org/10.3390/su14052897

Ribeiro, N., Gupta, M., Gomes, D., & Alexandre, N. (2020). Impact of PsyCap on affective commitment: Mediating role of affective well-being. *International Journal of Organizational Analysis, 29*(4), 1015–1029. https://doi.org/10.1108/IJOA-04-2020-2122

Saeed, B. B., Afsar, B., Hafeez, S., Khan, I., Tahir, M., & Afridi, M. A. (2018). Promoting employee's proenvironmental behavior through green human resource management practices. *Corporate Social Responsibility and Environmental Management, 26*(2), 424–438. https://doi.org/10.1002/csr.1694

Semedo, S., Coelho, A., & Ribeiro, N. (2018). The relationship between authentic leaders and employees´ creativity. What is the role of affective commitment and job resourcefulness? *International Journal of Workplace Health Management, 11*(2), 58–73. https://doi.org/10.1108/IJWHM-06-2017-0048.

Singh, S. K., Chen, J., Del Giudice, M., & El-Kassar, A. N. (2019). Environmental ethics, environmental performance, and competitive advantage: Role of environmental training. *Technological Forecasting and Social Change, 146*, 203–211. https://doi.org/10.1016/j.techfore.2019.05.032

Tajfel, H., & Turner, J. C. (1985). The social identity theory of intergroup behavior. *Political Psychology*, 7–24.

Tang, G., Chen, Y., Jiang, Y., Paille, P., & Jia, J. (2017). Green human resource management practices: Scale development and validity. *Asia Pacific Journal of Human Resources, 56*(1), 31–55. https://doi.org/10.1111/1744-7941.12147

Yücel, I., Ribeiro, N., & Gomes, D. (2020). Perceived organisational support and employees' performance: The mediating role of affective commitment. *International Journal Management and Enterprise Development, 19*(3), 187–202. https://doi.org/10.1504/IJMED.2020.108713

Zeng, Q., van Blokland, W. B., Santema, S., & Lodewijks, G. (2020). A company performance index for motor vehicle manufacturers: Company performance measurement with environmental concerns. *International Journal of Productivity and Performance Management, 70*(7), 1604–1628. https://doi.org/10.1108/IJPPM-08-2019-0393

Zhu, J., Tang, W., Wang, H., & Chen, Y. (2021). The influence of green human resource management on employee green behavior—A study on the mediating effect of environmental belief and green organizational identity. *Sustainability, 13*(8), 4544. https://doi.org/10.3390/su13084544

The Design of the Cyberphysical Navigator Model as a Sustainable Framework Enhancing Organisational Resilience

Qeis Kamran and Marcus Becker

Abstract Global societal disruptions, such as the 2008 financial crisis and the ongoing COVID-19 pandemic, introduce discontinuity for organizations. In the contemporary turbulent post-COVID-19 society, there is a critical need to establish resilience within the dynamics of organizational structures. Concurrently, advancements in technology and the pervasive connectivity-driven digitalization have transformed the organizational environment into a converged cyber-physical reality, blurring the boundaries between the digital and physical worlds. In this new reality, the management of internal and external communication becomes crucial for organizations to ensure resilience and agility. This context presents a unique opportunity intertwined with a considerable degree of complexity, thereby necessitating a new organizational adaptation. To address these challenges, this paper proposes the Cyber-physical Navigator Model (CNM) as a comprehensive framework that integrates augmented and virtual reality, amplified by artificial intelligence-driven man–machine interfaces, with the fields of cybernetics, design thinking, and new work. The CNM aims to support organizations in navigating through complex times, offering a more holistic approach to marketing management.

Keywords Complexity · Man–machine-interface · AR/VR · Cybernetics · Marketing strategy · AI · Design thinking · New work

Q. Kamran (✉) · M. Becker
International School of Management (ISM), Dortmund, Germany
e-mail: qeis.kamran@ism.de

Q. Kamran
Department of Engineering Technology, University of Twente, Enschede, The Netherlands

Weiden Business School, Ostbayerische Technische Hochschule Amberg-Weiden, Weiden in der Oberpfalz, Germany

C. Machado (ed.), *Building the Future with Human Resource Management*,
Management and Industrial Engineering, https://doi.org/10.1007/978-3-031-52811-8_4

1 Introduction

The primary objective of organizations worldwide has always been to navigate the complexities of their respective environments and establish a strong position. This resilient position is achieved through effective management, innovation capabilities, and operational excellence (Porter, 1985; Beckman & Barry, 2007). However, unforeseen disruptions such as the 2008 financial crisis and the current COVID-19 pandemic present challenges that are difficult to overcome. They highlight the limitations of profit-oriented and shareholder value-driven frameworks that fail to support managers in absorbing disruptive external shocks (Malik, 2011). The pursuit of organizational resilience and growth within an industry does not easily align on the same foundation, as the former addresses short-term industrial competition while the latter establishes long-term organizational viability within a different strategic context.

Moreover, the fields of marketing and management are undergoing transformations. The contemporary shift towards sustainability as a Zeitgeist compels managers to go beyond the traditional understanding of marketing based solely on economic theory and embrace a more holistic worldview. Con-currently, technological developments and digitalization are reshaping the organizational landscape, leading to a cyber-physical reality where the boundaries between the digital and physical worlds are increasingly blurred. This transformation significantly impacts how companies conduct business and how they align people through the emergence of ubiquitous global communication technologies.

While this presents a unique opportunity, it also entails a high degree of complexity, requiring organizational adaptation. Furthermore, the vast amount of data available to humanity has traditionally been confined to two-dimensional screens, limiting our ability to fully perceive and process information from the three-dimensional real world. Augmented and virtual reality (AR/VR) technologies address this limitation by bridging the gap between limited visualization capabilities and the massive amount of data to be observed and processed (Porter & Heppelmann, 2017).

In light of these considerations, scholars and practitioners must address the following questions:

1. How can a holistic model of man–machine interface support organizations in maintaining resilience within the complex and dualistic cyber-physical environment of a post-COVID society?
2. Which frameworks and principles should be integrated into the organizational structure to ensure an effective interplay between technologies and value-creating processes while simultaneously improving employee well-being?

This research paper aims to answer these questions by proposing the cyber-physical Navigator Model (CNM) as a comprehensive framework to guide managers in the turbulent and complex business landscape. The CNM integrates the aspects of AR/VR/AI technologies, design thinking principles, management cybernetics, and new work approaches to provide a holistic strategic framework for marketing operations.

2 The Complexity of Global Crises and Their Impact on Marketing and the Resilience of Organizations

Marketing, as a field of research, has always been characterized by its dynamic nature, influenced by a multitude of factors stemming from diverse sources. Over time, four key aspects have emerged as pivotal challenges for marketers and marketing firms. These aspects are depicted in Fig. 1 and have become crucial considerations for designing robust and effective frameworks and strategies that promote sustainable growth and ensure organizational homeostasis (Kamran, 2021).

The first significant trend, which is particularly relevant to this paper, is the increasing importance of new technologies such as augmented reality (AR), virtual reality (VR), and artificial intelligence (AI). In today's cyberphysical reality, organizations need to possess sufficient digital maturity to develop effective digital strategies.

The second trend is driven by global developments, including climate change and environmental pollution, which have heightened societal demands for sustainability. Organizations must incorporate sustainable practices into their operations to succeed with a strategic foresight.

The third trend involves the need for quick and effective responses to emerging complexities through the lens of cybernetics. This is exemplified by the case of Volkswagen AG, which demonstrated incompetence in effectively managing the Dieselgate scandal.

The final trend identified is the inefficiency of traditional marketing approaches in the complex reality of interconnected global value chains (GVC). The globalization of value chains and markets has increased the influence of global disruptions on organizational resilience. Wars, pandemics, and financial crises can significantly impact organizations' ability to adapt.

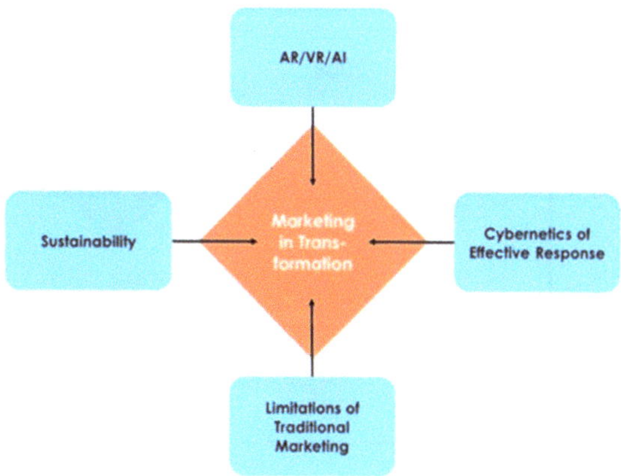

Fig. 1 Transformation of marketing in 21st century (Kamran, 2021)

Figure 2 illustrates the profound influence of two major crises—the global financial crisis of 2008 and the ongoing COVID-19 pandemic—on organizations worldwide. These crises have exposed the need for change and adaptation within the current system of global markets and the nature of capitalism. However, many companies have failed to adjust their strategies and attitudes toward sustainable economic, social, and intellectual issues. Short-term profit maximization has overshadowed considerations of organizational viability, environmental sustainability, and social responsibility (Hall, 1962).

To foster resilience and innovation within organizations, a shift in mindset is necessary. Traditional organizational structures focused on shareholder value, and fragmented marketing approaches are inadequate in turbulent environments. The Volkswagen Dieselgate scandal exemplifies the consequences of an absence of resilient leadership and a flawed understanding of foreign markets. Embracing the concept of shared value allows organizations to establish long-term resilience by acknowledging the importance of responsive and sustainable actions within a socio-environmental context, which ultimately leads to increased profits in the long run.

The current COVID-19 pandemic has had a significant impact on various industries, surpassing the effects of the 2008 crisis. The recovery process for firms will likely be prolonged, emphasizing the need for adaptation and the development of innovative and resilient business frameworks (Davies, 2020). Marketing plays a

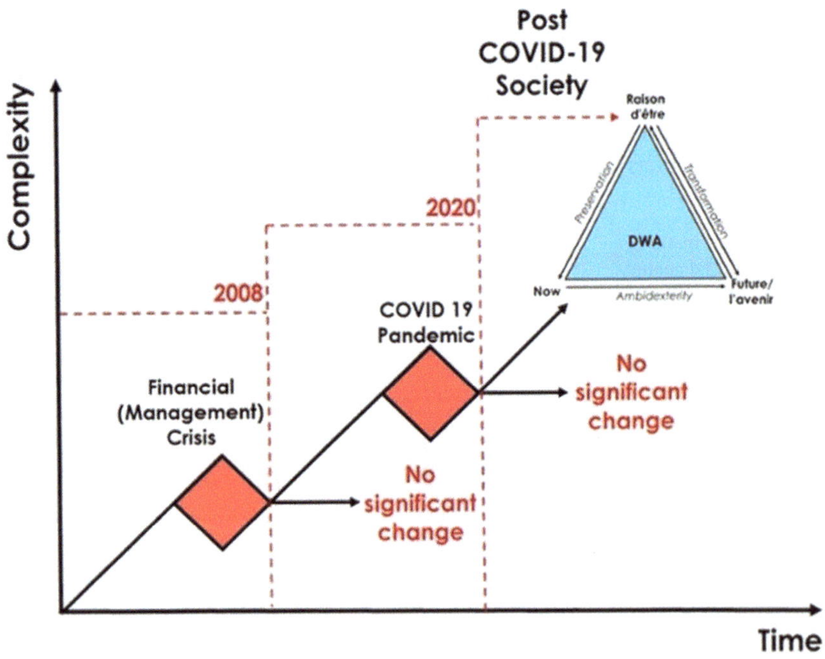

Fig. 2 Complexity of the twenty-first century and the DWA model (Author´s own creation)

crucial role in this endeavor by creating organizational structures capable of navigating the crisis and by highlighting the importance of aligning technologies and innovative work concepts. Remote working and efficient home-office infrastructures have become central success factors for organizational resilience, yet many companies still lack the necessary competence in this essential aspect.

The model presented above highlights the significant impact of the financial crisis and the subsequent COVID-19 pandemic on organizations. The lack of change and adaptation following the financial crisis left organizations ill-prepared to handle the complexities and disruptions caused by the pandemic. However, the severe consequences of the pandemic have forced organizations to reevaluate their strategies and values, leading to a paradigm shift in thinking.

To prevent a situation where organizations resist change, a theoretical framework called the Design Weltanschauung Model (DWA) has been proposed. The DWA aims to support organizations in navigating through turbulent times and coping with the wide variety of complexities present in the markets. It serves as a framework to align the essential pillars of an organization that are crucial for its resilience in a world of global disruptions.

To introduce the DWA, it is important to provide a comprehensive understanding of the framework. Additionally, a brief introduction to the concept of design is necessary. Design, in this context, refers to a strategic approach that goes beyond aesthetics and encompasses the intentional shaping of products, services, and systems to meet specific goals and solve complex problems.

By adopting a design-oriented perspective, organizations can foster resilience, innovation, and adaptability. The DWA framework offers a structured approach to incorporating design principles and practices into an organization's strategic decision-making processes. It provides guidance on how to align various aspects of the organization, such as leadership, culture, processes, and communication, to create a cohesive and resilient system.

Overall, the DWA framework aims to facilitate organizational transformation and enable organizations to thrive in the face of disruptions by embracing a design-oriented mindset and approach.

2.1 Brief History of Design

The concept of design and its associated methodologies, including design thinking, have a longstanding history that extends over six decades. The emergence of these approaches can be traced back to the 1950 and 1960s when designers sought to address the complex challenges that arose in the aftermath of the Second World War. These new design methods, rooted in the principles of creativity, aimed to incorporate emerging technologies and their application for human use.

During this period, designers began to recognize that their approaches to integrating new technologies were less articulate and comprehensible compared to those employed by scientists and engineers responsible for their development. This realization prompted the organization of conferences on design methodology and the publication of books that delved into the subject. Additionally, the establishment of research-based journals such as "Design Studies," "Design Issues," "Design History," "Research in Engineering Design," and "Languages of Design" further contributed to the proliferation of design as a distinct research field.

A pivotal figure in promoting design as a cognitive framework was Herbert Simon, a renowned cognitive scientist and recipient of the Nobel Prize in economics. Simon posited in 1969 that design thinking was not confined solely to professional engineers but encompassed anyone who engaged in the process of devising courses of action aimed at transforming existing situations into preferred ones (Fig. 3).

Building upon Simon's foundation, Bruce Archer emerged as a prominent advocate for recognizing design as a third paradigm of education, alongside science and humanities. Archer argued that both science and humanities possessed inherent limitations, neglecting certain crucial aspects. Humanities, for instance, focused on human expressions and values but disregarded the practical and creative elements associated with the fine arts. Science, on the other hand, relied on observation-based approaches and notations to generate theories and knowledge, often neglecting the practical implementation and experiential dimensions of design. Archer proposed that design, with its emphasis on the practical and creative aspects, should be acknowledged as a distinct academic discipline.

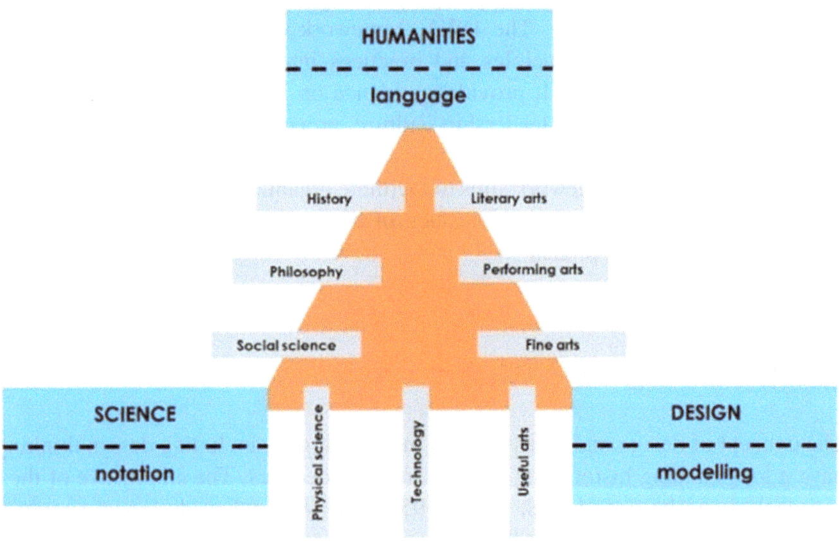

Fig. 3 Relationship between humanities, science and design (Author´s own creation based on Archer, 1965, 1979)

Archer's proposition emphasized the significance of integrating creative and practical problem-solving approaches into research and education, leading to the development and widespread adoption of design thinking as a problem-solving framework in various fields, including management. This recognition facilitated the incorporation of design principles into managerial practices, enabling organizations to tackle complex challenges and foster innovation.

"The natural sciences are concerned with how things are. Ordinary systems of logic the standard propositional and predicate calculi, say serve these sciences well...Design, on the other hand, is concerned with how things ought to be, with devising artifacts to attain goals" (Simon, 1969). The field of design is characterized by its focus on the act of creation and the production of tangible artifacts, aspects that are often overlooked by the aforementioned disciplines. Unlike the humanities and sciences, design encompasses the realms of making and doing, where the language of modeling plays a central role. This language of design can be expressed through various means, such as sketching and drawing, allowing designers to convey their ideas and concepts (Archer, 1979).

Cross (1999) recognized the potential of incorporating virtual reality (VR) into the language of design, particularly with the advent of advanced computing technologies. VR has emerged as a powerful tool that enables designers to visualize and communicate their ideas in an immersive and interactive manner. By leveraging VR, designers can enhance their ability to model and prototype new concepts, fostering a more dynamic and engaging design process. This integration of VR into the language of design signifies its growing significance and potential impact on the field.

2.2 Design Thinking—The Managerial Approach of Designerly Application

In recent years, design thinking has gained significant traction as a prominent approach in various fields, particularly in management. It represents a practical manifestation of design principles and methodologies, derived from the broader domain of design science research. Design thinking is distinct from its academic counterpart known as "designerly thinking," which pertains to the scholarly exploration of the practical skills and competencies exhibited by professional designers. The academic discourse of designerly thinking involves the formulation of theories that aim to interpret and characterize the non-verbal expertise of designers. On the other hand, design thinking, or DT, is a simplified form of the design discourse that is accessible to individuals without a scholarly background, serving as a practical framework and methodology, particularly in management contexts (Cross, 1982, 1993; Johansson-Sköldberg et al., 2013).

Hassi and Laakso provided a comprehensive overview of design thinking by highlighting its key elements and characteristics, which can be extrapolated from Fig. 4. This modern interpretation of design thinking encompasses three fundamental

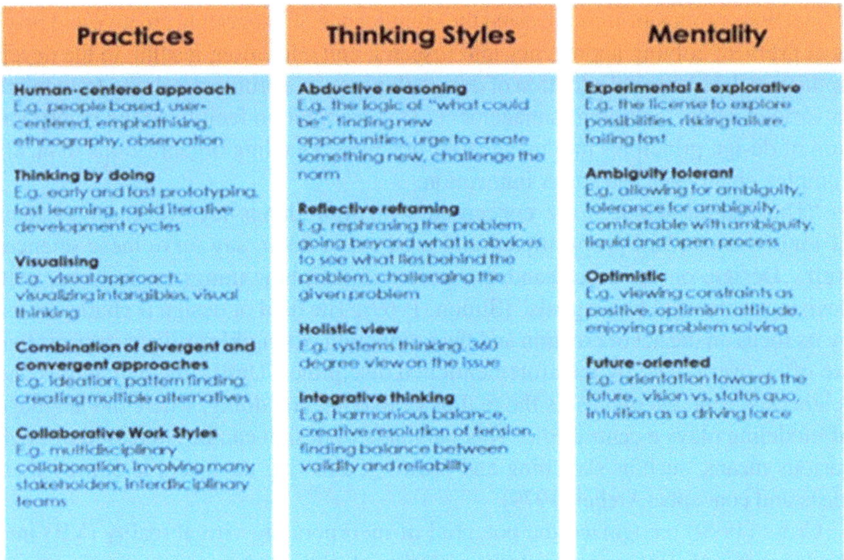

Fig. 4 Elements of modern DT. *Source* Authors' own creation based on Hassi and Laakso (2011)

aspects: practices, cognitive approaches, and mindset. Notably, even in this contemporary understanding, design thinking incorporates elements such as the utilization of visualizations as a language and the adoption of a holistic and interdisciplinary approach. These aspects have deep historical roots in the evolution of design thinking and have persisted throughout its development (Hassi & Laakso, 2011).

2.3 The DWA Model— Model of Rapprochement Between Design and Marketing

The proposed Design Weltanschauung (DWA) model, depicted in Fig. 5, builds upon the foundational principles of design as articulated by Archer (1965, 1979) and Simon (1969), as well as the contemporary understanding of design thinking outlined by Johansson-Sköldberg et al. (2013). The DWA model aims to provide organizations with a framework to effectively navigate and respond to the turbulence and complexities present in the market, including challenges posed by events such as the COVID-19 pandemic or the financial crisis. Thus, as Alexander (1964) coined patterns, pivotal to the process of creating form. Recognizing the limitations of traditional marketing approaches in fostering resilience and addressing the multifaceted nature of contemporary marketing trends, the authors advocate for the incorporation of a marketing logic grounded in Archer's concept of design as a third pillar of human knowledge alongside science and humanities. The DWA model represents an applied

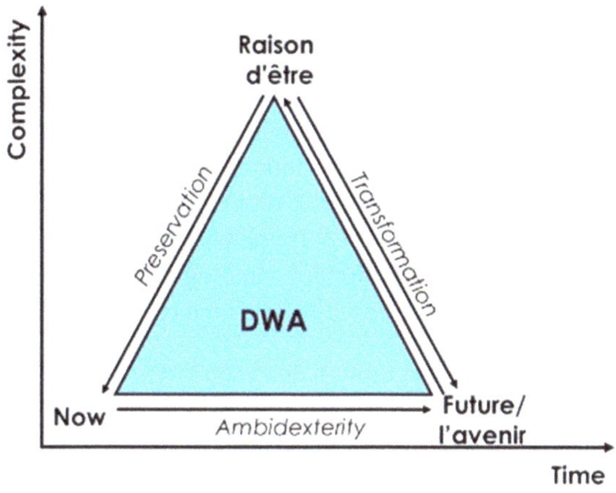

Fig. 5 DWA model (Kamran, 2021)

framework for marketing that offers managers a holistic foundation, encompassing various disciplines such as philosophy, cybernetics, aesthetics, and others, within a "designerly context" that enables them to observe and leverage potential synergies within the global market (Kamran, 2021).

"A well-defined GVC relates the firm to its wider environment's changing characteristics, while it understands the preserving and evolving nature of the firms' and GVC's raison d'être" (Kamran, 2021). The DWA framework is built upon three key dimensions: ambidexterity, transformation, and preservation. Ambidexterity refers to an organization's ability to effectively manage current operations and processes while simultaneously exploring new opportunities and markets (Tushman & O'Reilly, 1996). Given the dynamic and crisis-prone nature of the market, it is essential for organizations to design innovative responses to disruptions while maintaining operational excellence in core processes.

The DWA framework addresses three fundamental questions crucial for organizational success and resilience:

1. Who are we, and what is our purpose? This pertains to understanding the organization's identity, legitimacy, and overarching objectives.
2. What needs to be done today? This focuses on managing existing priorities within the organization's current capabilities and timeframe.
3. What needs to be done in the future, and how do we initiate the transformation process today? This highlights the importance of anticipating future needs and proactively undertaking transformative actions.

By adopting the DWA framework, managers gain a robust theoretical foundation that can enhance their ability to lead organizations in the post-COVID society. The framework's three pillars ensure operational excellence, efficient responses to diverse challenges, and a long-term perspective that goes beyond short-term profit maximization and silo thinking.

Furthermore, it is important to acknowledge the significance of digitalization and the rise of technologies such as augmented reality (AR), virtual reality (VR), and artificial intelligence (AI). These technologies are not merely abstract concepts but are already shaping our current reality. The convergence of physical and virtual worlds is blurring the boundaries, and in the post-COVID-19 society, this phenomenon will be further amplified by the exponential potential for scaling within digital spaces.

In summary, the DWA framework, coupled with an understanding of the transformative impact of digital technologies, provides organizations with the necessary tools and perspectives to thrive in an increasingly complex and interconnected world.

3 Creating a Theoretical Framework to Cope with a Cyber-Physical Post-COVID Society

A wider model that establishes a relevant framework to enhance organizational resilience within the post-COVID-19 society and to act intelligently in the cyber-physical reality is introduced here. It is incorporating the DWA model and expanding its logic with aspects of AR/VR man–machine interface and the integration of essential components of new work. Furthermore, the notion of design thinking (DT) and the derived process are a central feature of this holistic approach aimed to enhance organizational creativity and establish more adequate solution-search-spaces for an adequate response.

The framework constructs additionally on the principles of cybernetics (Beer, 1959, 1995; Wiener, 1948), in particular Ashby's (1956) law of requisite variety. Ashby's law implies that only variety can absorb variety. In this case, variety is the measure of complexity that can only be effectively managed by an equivalent amount of probabilistic complexity within the structural dynamics of the controller model (Christopher, 2007).

3.1 Cyber-Physical Reality as the Embedded Environment the Model

The notion of cyber-physical reality, depicted in Fig. 6, represents the convergence of the physical and digital worlds into a new unified reality. This convergence is a defining feature of the complex nature of the twenty-first century. In this global society, there exists a web of interconnected and ubiquitous networks involving various actors. Connectivity is a fundamental aspect, and it has given rise to a global converging culture where shared values and concerns play a guiding role (Kamran, 2021).

Moreover, in today's digital age, an organization's digital presence has become essential. If an organization cannot be found on platforms like Google, it raises doubts

Fig. 6 Cyber-physical environment (Kamran, 2021)

not only about its digital existence but also about its physical presence in the real world. The digital realm has become intertwined with the physical, and organizations need to establish a strong online presence to maintain their relevance and visibility.

This highlights the significance of embracing digital technologies and understanding the interconnectedness of the physical and virtual realms. It is crucial for organizations to adapt to the evolving cyberphysical reality and leverage digital platforms to establish their presence and engage with their target audience effectively.

3.2 Organisational Unit

In order to navigate through the disruptions caused by the COVID-19 pandemic and the complexities of the cyber-physical environment, the authors introduce a viable organizational structure within the framework of a control system, as depicted in Fig. 7. This structure aims to address the following key aspects: the integration of

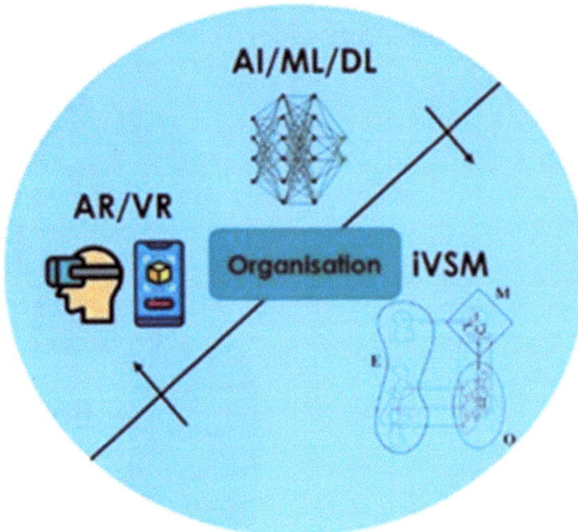

Fig. 7 Organisational unit of the model (Authors' own creation)

cutting-edge technologies such as augmented reality (AR), virtual reality (VR), and artificial intelligence (AI), and the establishment of a functional interplay between humans and machines.

To achieve organizational viability, it is crucial to have a comprehensive understanding of the organization's structure. For this purpose, the authors propose the adoption of the Viable System Model (VSM), which provides a suitable framework. Building upon the components of the man–machine interface, namely AR/VR, combined with the capabilities of AI and machine learning, and supported by the robustness of a cybernetically established management model, the authors introduce an extended version of the model called the iVSM (integrated Viable System Model) model.

The iVSM model incorporates the latest technologies and leverages the synergies between human intelligence and machine capabilities. This integration enables organizations to adapt and thrive in the cyber-physical reality by effectively managing complexities, optimizing decision-making processes, and ensuring organizational resilience.

By embracing the iVSM model, organizations can harness the power of AR, VR, and AI to create a dynamic and responsive organizational structure that is capable of navigating through turbulent times, addressing emerging challenges, and seizing new opportunities in the digital age

3.2.1 AR/VR/AI Man–Machine Interface

Figure 8 illustrates the importance of a well-functioning augmented reality (AR) and virtual reality (VR) man–machine interface as a crucial component of the control system. This interface plays a vital role in achieving optimal cyber-physical

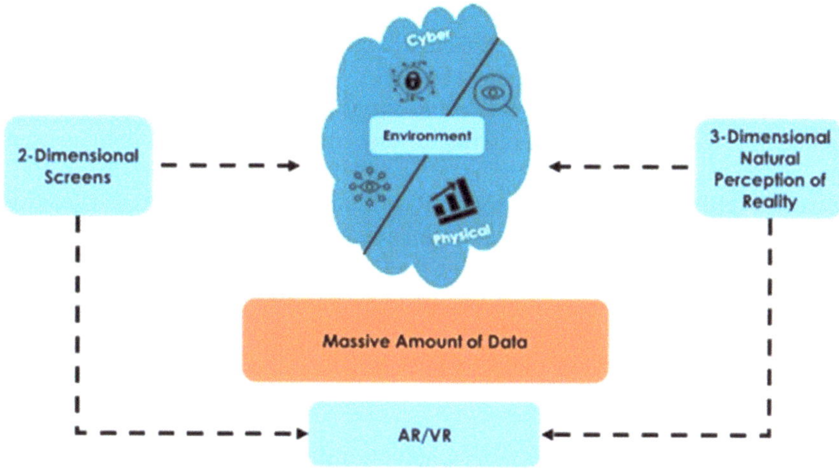

Fig. 8 AR/VR and the cyberphysical environment (Authors' own creation)

resilience and operational excellence within the organization. It addresses the challenge posed by the limited visualization capabilities of 2-dimensional screens in effectively processing the vast amount of data originating from the cyber-physical environment.

The integration of AR and VR technologies into the organizational structure helps bridge the gap between traditional screens and the complex data that needs to be observed and efficiently processed. AR enhances the real-world environment by overlaying digital information, while VR creates immersive simulated experiences. Together, they provide a more comprehensive and immersive way to interact with and understand the cyber-physical environment.

By incorporating AR/VR into the organizational structure, organizations can enhance their ability to visualize and comprehend complex data, enabling more informed decision-making and improving operational efficiency. This integration empowers individuals within the organization to navigate and interact with the cyber-physical reality in a more intuitive and immersive manner, leading to improved problem-solving, innovation, and overall organizational performance.

The integration of augmented reality (AR) and virtual reality (VR) technologies into an organization can be achieved through various measures across different activities within the value chain. The COVID-19 pandemic has heightened the need for remote working and communication, prompting companies to adopt remote communication platforms such as Zoom, Microsoft Teams, or Google Hangouts. As remote work and teaching become increasingly prevalent in the emerging cyber-physical post-COVID society, further research is warranted to explore the integration of AR/VR, enabling a more immersive remote collaboration experience. Platforms like Virbela, a virtual campus application, immerse users in a virtual version of their organizations or universities, facilitating group work and collaborative activities.

Enhancing such platforms with AR/VR technology presents a compelling alternative to traditional software like Zoom (Baker, 2020).

Immersive technology, exemplified by IKEA's Place AR-app (Ozturkcan, 2020), also holds significant potential in marketing campaigns. Smartphone-based AR applications have gained popularity, offering diverse implementation possibilities along the customer journey and serving as a valuable tool for omnichannel marketing strategies (Hilken et al., 2018). With VR devices becoming more accessible to a wider audience, organizations need to explore how VR can be integrated into their marketing strategies. Particularly during the COVID-19 pandemic, a well-designed VR strategy can facilitate communication between firms and potential customers, serving as a promising alternative to physical store visits (Barnes, 2016). Additionally, AR/VR technology can significantly impact logistics, customer service, and employee training, adding a new dimension to the functions of an intelligent organization.

VR has the ability to immerse users in entirely different virtual worlds and settings, making it particularly valuable for education in hazardous environments or professions where mistakes can have severe consequences. VR provides a risk-free environment for participants to learn and practice tasks without significant pressure. This aligns with Edgar Dale's "Cone of Experience" model, which suggests that people retain 90% of what they do compared to 10% of what they read (Dale, 1969). Training programs for firefighters and medical professionals, such as surgeries, have successfully utilized VR technology to enhance learning outcomes(Sutherland et al., 2019).

The integration of VR into employee training has demonstrated benefits beyond extreme environments. A notable example is Walmart, which introduced tethered VR technology at its Walmart Academies across the United States. Walmart developed over 45 training modules using VR to teach employees specific situations, such as the introduction of new technologies or handling customer interactions during busy periods like Black Friday. Participants using VR reported higher training satisfaction, with 70% of employees trained with VR outperforming those using traditional techniques (Carruth, 2017). The AR/VR man–machine interface aims to encompass all processes within the organizational context (Porter & Heppelmann, 2017).

To complement the AR/VR interface, the integration of artificial intelligence (AI), machine learning (ML), and deep learning (DL) into the corporate structure is crucial. These technologies support the AR/VR interface and enable organizations to efficiently cope with large amounts of data. Additionally, AI plays a central role in detecting hidden patterns emerging from the convergence of cyber-physical realities. In this context, the Artificial General Marketing Intelligence (AGMI) framework has been proposed as a suitable theoretical framework for successful AI integration, providing guidance on how an AI-amplified embodiment can be prototyped and tested (Kamran, 2021).

3.2.2 AR/VR Integration and Employee/student Wellbeing

In addition to its role in creating organizational resilience, there is a growing interest in using immersive experiences to improve the health and wellbeing of employees. Research has shown that VR content used for stress reduction can have positive effects on both qualitative feedback and physical parameters like heart rate (Naylor et al., 2019). This makes VR applications particularly relevant in the context of the COVID-19 pandemic. For example, organizations can provide VR nature experiences during breaks or offer meditation courses that include breath exercises (Ahmaniemi et al., 2017; Naylor et al., 2019).

These measures can also be applied to enhance psychological wellbeing during the COVID-19 crisis. In addition to the stress-reducing VR experiences mentioned earlier, organizations can implement VR classes to support employees with mental health issues. Private sessions with professional therapists using VR technology are also possible interventions (Frewen et al., 2020). Furthermore, the integration of platforms and the gamification of remote working and learning experiences can bring a sense of novelty and excitement to individuals who may be experiencing isolation and anxiety in current times. Additionally, VR can be utilized to improve workplace design, which has a significant impact on employee wellbeing (Michalos et al., 2018).

3.2.3 The Viable System Model as a Solid Foundation for the Organisational Structure

The field of management cybernetics, pioneered by Beer (1959, 1979, 1984, 1995), holds a central position in the overall structure and functioning of the depicted control system. The Viable System Model (VSM), derived from principles of bionics and cybernetics, serves as the most advanced organizational model, drawing inspiration from the brain and the human nervous system (Beer, 1995).

The VSM consists of five interrelated systems, each fulfilling specific functions within the organization:

1. System 1 represents the operative and value-creating activities that form the core processes defining the organization's legitimacy.
2. System 2 encompasses the measures responsible for coordinating the autonomously functioning System 1 units.
3. System 3 is responsible for maintaining and optimizing the stability of the interactions be-tween the different System 1 units. It serves as a crucial link between operational activities and top management.
4. System 4 assumes a more strategic role, dealing with future-oriented considerations and the uncertainties associated with them.
5. System 5 acts as a mediator to mitigate potential conflicts between System 3 and System 4. It represents the norms, values, and ethos upon which the organization

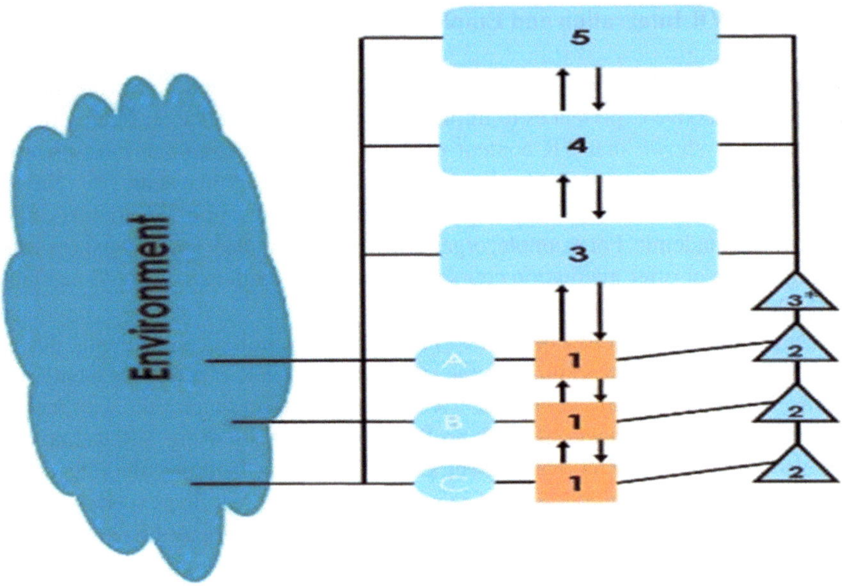

Fig. 9 Simplified sketch of the VSM (Author's own creation based on Beer, 1995; Kamran, 2021: 293)

is built and is embodied by the executive management and leadership (Beer, 1984).

By adopting the VSM framework, organizations can establish a coherent structure that effectively addresses the diverse operational and strategic requirements while maintaining overall stability and adaptability (Fig. 9).

3.3 *Management Unit*

Indeed, the implementation of the measures outlined above, such as the integration of new technologies or the reorganization of business units, will necessitate substantial changes within organizational structures. To effectively navigate these changes and ensure successful adaptation to the cyberphysical environment, a strong management unit is essential (Fig. 10). This management unit should be built upon inspiring and innovative principles of working and leading employees.

The management unit plays a pivotal role in driving and managing organizational transformation. It should possess the skills, knowledge, and mindset necessary to navigate the complexities and challenges associated with the integration of new technologies, the creation of viable structures, and the overall adaptation to the cyber-physical environment. This unit should foster a culture of innovation, collaboration,

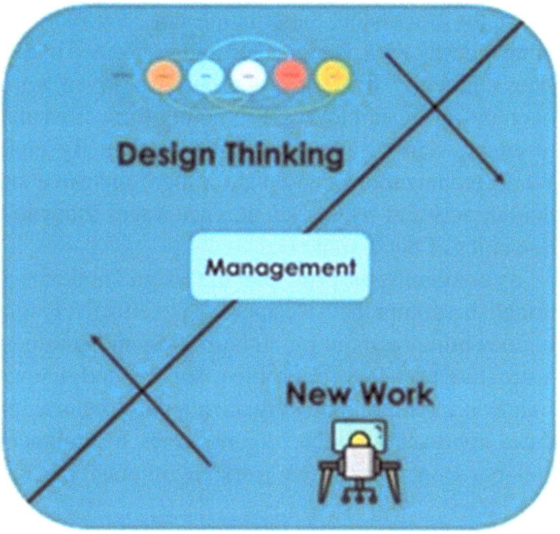

Fig. 10 Management unit of the model (Author's own creation)

and continuous learning to enable employees to embrace change and contribute to the organization's success.

Effective leadership within the management unit involves inspiring employees, providing clear direction and vision, and empowering them to take ownership of their roles in the organizational transformation. Leaders should encourage a culture of creativity, experimentation, and risk-taking while ensuring that employees feel supported and motivated throughout the change process.

By establishing a strong management unit that embraces innovative principles and empowers employees, organizations can enhance their capacity to adapt to the demands of the cyberphysical environment and successfully navigate the complexities of change.

The introduction of Design Thinking (DT) as a process system of effective management and the principles of New Work are two complementary approaches that contribute to creating a better organizational culture and successfully absorbing the changes brought about by digitalization and automation.

Design Thinking, with its human-centered and iterative problem-solving approach, emphasizes empathy, collaboration, and experimentation. It encourages organizations to understand the needs and perspectives of their stakeholders, including customers, employees, and partners, and to design innovative solutions that address those needs. By adopting a DT mindset, organizations can foster a culture of creativity, adaptability, and continuous improvement. This approach is particularly valuable in the context of digitalization and automation, as it encourages organizations to explore new possibilities, challenge existing assumptions, and leverage technology to create meaningful and user-centric experiences.

On the other hand, the principles of New Work provide a framework for reimagining the way work is organized and performed in the digital age. New Work emphasizes autonomy, flexibility, and purpose-driven work. It recognizes the importance of empowering employees, promoting work-life balance, and leveraging technology to enable remote work and collaboration. By embracing the principles of New Work, organizations can create a more inclusive and engaging work environment that attracts and retains talent, encourages innovation, and adapts to the changing dynamics of the digital era.

By combining the principles of Design Thinking and New Work, organizations can establish a comprehensive approach to effectively manage the challenges and opportunities brought about by digitalization and automation. Design Thinking provides a structured process for problem-solving and innovation, while New Work offers a framework for rethinking organizational structures, processes, and culture. Together, these approaches enable organizations to embrace change, foster creativity and collaboration, and create a work environment that supports continuous learning and growth.

3.3.1 Design Thinking Process

Design Thinking (DT) is a managerial approach that offers managers a fundamental theoretical framework for transforming the designerly understanding of doing things and cultivating a spirit of innovation within organizations (Archer, 1979; Cross, 1982; Simon, 1969). DT combines a corresponding mindset, an appropriate toolbox, and a range of processes to facilitate effective organizational creative dynamics (Brenner & Uebernickel, 2016).

DT has been interpreted and applied in various ways by different scholars and practitioners. For example, Lawson (2006), IDEO (Brown, 2008), and the Hasso Plattner Institute (HPI) at the University of Potsdam (Plattner et al., 2009) have all contributed to the evolution of DT. In this paper, the authors have integrated the holistic framework developed by the HPI model.

The Fig. 11 illustrates the characteristics of modern DT, organized into categories that encompass the relevant practices, thinking styles, and mentality associated with this approach (Hassi & Laakso, 2011).

The Fig. 11 illustrates the characteristics of modern Design Thinking (DT). It is divided into three categories: Practices, Thinking Styles, and Mentality. Under Practices, it includes elements such as empathy, ideation, prototyping, and testing. Thinking Styles encompass characteristics like divergent thinking, reframing problems, and integrative thinking. Lastly, mentality includes attributes such as curiosity, optimism, and resilience.

By adopting DT principles and practices, managers can foster a culture of innovation, collaboration, and problem-solving within their organizations. This approach emphasizes the importance of under-standing user needs, challenging assumptions, and iterating on solutions through prototyping and testing. It encourages managers

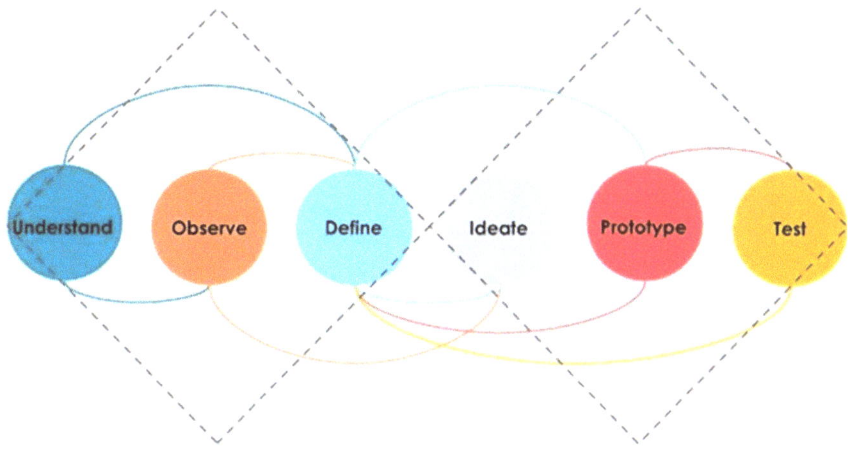

Fig. 11 HPI design thinking process (Authors' own creation based on Plattner, Meinel, & Weinberg, 2009 and Tschepe, 2017)

to adopt a mindset of curiosity, optimism, and resilience, which are crucial for navigating the complexities of the digital age.

Overall, DT provides a valuable framework for managers to effectively implement and integrate in-novation and creativity into their organizational processes. By embracing the principles of DT, managers can drive positive change, create meaningful experiences for their stakeholders, and adapt to the evolving demands of the digital era.

The illustration (Fig. 11) presents a visual representation of the six stages of the Design Thinking (DT) process, with each stage depicted by a circle. The diagram also illustrates the connections between the stages and indicates where iterations should occur (Plattner et al., 2009). The two dashed diamonds represent the distinct phases of the model: divergent thinking and convergent thinking.

Divergent thinking, which is the first phase, focuses on the problem space. It involves the identification and exploration of significant problems or user needs. This stage aims to gather a wide range of perspectives, insights, and potential solutions to generate innovative ideas. It encourages open-mindedness, creativity, and the generation of diverse options (Kamran, 2021; Tschepe, 2017) (Fig. 12).

Convergent thinking, on the other hand, deals with the solution space of the DT model. This phase involves analyzing and evaluating the ideas generated during the divergent thinking phase. It requires making decisions, narrowing down options, and selecting the most promising ideas to move forward. Convergent thinking involves critical evaluation, feasibility assessment, and the synthesis of ideas into actionable solutions (Kamran, 2021; Tschepe, 2017).

Successful DT implementation in the management discourse often relies on a sophisticated combination of convergent and divergent thinking approaches within multidisciplinary teams. This collaborative and interdisciplinary approach allows for

Fig. 12 New Work Charta (Authors' own creation based on Väth, Gstöttner & Soballa, 2019)

the exploration of diverse perspectives, the synthesis of ideas, and the development of innovative solutions (Hassi & Laakso, 2011).

The Design Thinking (DT) process consists of six distinct phases, each serving a specific purpose in the development of innovative solutions:

1. Understand: The first phase involves introducing an interdisciplinary team to the problem at hand and gaining a broader context of the issue. The team aims to have an objective and open mindset to avoid misunderstandings and gain a comprehensive view of the problem's complexity. Activities during this phase may include gathering information, conducting interviews, and getting a first contact with the target group or users.
2. Observe: In the second phase, the focus shifts to the target group or users related to the problem. The goal is to deeply understand their needs, values, and behaviors. This phase may involve further interviews and observations to gather insights into the users' perspectives and experiences.
3. Define: Building upon the insights gathered in the previous phases, the third step aims to create a well-crafted problem statement that incorporates all the characteristics and challenges of the identified problem. The problem statement also takes into account the special needs of the users.
4. Ideate: In the ideation phase, the thinking mode switches to a convergent state, and the team starts generating potential solutions for the problem. The focus is on

quantity over quality, encouraging a highly innovative and creative atmosphere within the interdisciplinary team. The goal is to generate a wide range of possible solutions, avoiding early judgment to foster creativity.

5. Prototype: The prototyping phase involves transforming the selected ideas into tangible representations or mock-ups. This can be done through physical objects, activities, or processes. The purpose of prototyping is to learn about new challenges and risks that may arise during implementation. Early and inexpensive prototypes are encouraged to enable fast iterations and modifications to arrive at more refined solutions.
6. Test: The last stage is all about testing the prototypes with both team members and end-users, such as employees in the case of implementing a new control system for the cyberphysical environment. The testing process is highly iterative, allowing developers to make modifications or eliminate prototypes based on feedback. Customer feedback is also embedded in this phase. The more advanced prototypes are tested in real-world contexts, leading to a narrowing down of viable solutions until the optimal one is identified.

Throughout the DT process, the focus on interdisciplinary collaboration, empathy for users, and iterative development allows organizations to arrive at innovative and user-centric solutions to complex problems. By following this iterative and human-centered DT process, organizations can foster innovation, address complex problems, and create meaningful solutions that meet the needs of users and stakeholders.

3.3.2 New Work New Culture

The theory of new work emphasizes the importance of adapting work structures and forms to thrive in a complex digital market and foster long-term success, innovation, freedom, and creativity within organizations. It recognizes that traditional work models may not be effective in the face of rapid technological advancements and changing market dynamics. Key proponents of the theory, such as Bergmann (1977, 2019) and Väth (2016), have contributed to its development. In addition we understand the essence of engineering designerly artifacts as a purposeful activity targeted towards fulfilling human aspirations, especially those which can be unconcealed by the technological dimensions and possibilities of cultural semantics pertaining to products' provisions (Asimov, 1962).

The core principles of the theory of new work involve reimagining work in ways that enhance employee motivation, engagement, and flexibility. It advocates for a shift from rigid hierarchical structures to more decentralized and autonomous work arrangements. This includes promoting self-management, employee empowerment, and a focus on individual and team autonomy. New work also emphasizes the importance of work-life balance, employee well-being, and meaningful work experiences.

Organizations that embrace the theory of new work aim to create a modern work culture that encourages collaboration, innovation, and continuous learning. They recognize the value of diverse perspectives and interdisciplinary teams, fostering an environment that enables creativity and problem-solving. New work also encourages the use of technology to support flexible work arrangements, remote collaboration, and digital communication tools.

By embracing the theory of new work, organizations can adapt to the demands of the digital market, remain competitive, and cultivate an environment that attracts and retains talented employees. It acknowledges the need for ongoing evolution and transformation in work structures and practices to ensure long-term success in an ever-changing business landscape

Frithjof Bergmann, the Founder of New Work

Frithjof Bergmann, an American social philosopher, is considered the founder of the New Work movement. His work revolves around the concepts of freedom, self-esteem, and redefining the relation-ship between people and work. Bergmann argued that the traditional wage labor system was outdated and advocated for an alternative model of work, which he called "new work."

According to Bergmann's vision, contemporary work should be divided into three equal parts:

1. **Reduced Wage Labor**: Bergmann proposed cutting down conventional wage labor to a minimum, reserving only one-third of people‘s time for traditional work. This would allow individuals to have more time for other pursuits.
2. **High-Tech Self-Provisioning (HTSP)**: The saved time from reduced wage labor should be used for work that individuals truly desire to perform. Bergmann envisioned a system of high-tech self-provisioning, where people fulfill their basic needs through technology and self-sufficiency. This aspect aims to satisfy individuals' needs, which previously were the main motivation for engaging in regular work.
3. **Autonomous Living**: The last aspect of Bergmann‘s vision pertains to the economic rationale of a highly technologized and autonomous way of living. After experiencing life without technological aids while living secluded in the woods of New Hampshire, Bergmann realized that lack of technology consumed significant time and effort for survival. He sought to free up time for more creative pursuits.

While Bergmann's ideas may be seen as radical by some, they have gained popularity in recent years as the term "new work" has become more widely used. However, it is important to note that implementing some of Bergmann's ideas, such as the abolition of paid labor, might encounter significant challenges and opposition. Nonetheless, his work has inspired researchers and organizations to explore and implement modernized versions of his ideas to create better workplaces for the future.

A Modern View on New Work

The concept of New Work, as originally proposed by Frithjof Bergmann, raises important questions for organizations in the context of the cyberphysical reality. It challenges the traditional understanding of work and calls for possible adaptations to the current work structure. However, in the modern context, the term "New Work" lacks a uniform framework and has sometimes been misused to describe minor improvements in working conditions, rather than addressing the more radical and critical topics proposed by Bergmann, such as the reduction of wage labor or a philosophical revaluation of work.

In today's complex and rapidly changing environment, especially in the aftermath of COVID-19, the term "New Work" has become somewhat of a buzzword, losing its original depth and scope. Many discussions around it tend to focus on incremental changes, efficiency improvements, and employee engagement, while overlooking the more profound and transformative aspects that may be necessary in the face of technological advancements and potential job eliminations.

To understand the true essence of New Work and its potential impact on organizations in the cyber-physical reality, a holistic view is essential. Modern New Work encompasses incremental changes across various dimensions, including psychological, social, technological, organizational, and political aspects. This broader perspective acknowledges that implementing New Work principles involves more than just increasing profits or employee loyalty. It requires considering the psychological well-being of employees, embracing new technologies, rethinking organizational structures, and addressing broader societal implications (Hackl et al., 2017; Helmold, 2021; Väth, 2016).

To provide guidance for organizations navigating this disruptive landscape, the authors of the paper suggest the New Work Charta (NWC) as a suitable tool. The NWC, created by the think tank 'Humanfy' led by Markus Väth, outlines five principles that organizations should follow to successfully implement New Work practices. These principles likely encompass the more holistic and transformative aspects of New Work, allowing organizations to embrace change and create a more innovative and adaptive work culture (Väth et al., 2019).

The author has made modifications to the social responsibility aspect of the New Work Charta (NWC) by incorporating the concept of shared value. This modification aims to inspire managers to adopt a more holistic and business-oriented perspective on corporate social responsibility, moving away from viewing it as a mere obligation to be fulfilled. The concept of shared value, as proposed by Porter and Kramer (2011), highlights the interconnectedness between social and economic goals, suggesting that organizations can create value for themselves while also addressing social and environ-mental challenges.

Outlined below are the five principles of the NWC, as presented by the authors:

1. **Self-determined work** (Freedom): This principle emphasizes the importance of autonomy and self-determination in the workplace. It encourages organizations to empower employees, al-lowing them to have a say in decision-making processes and providing opportunities for personal and professional development. Self-determined work fosters a sense of ownership and motivation among employees, leading to higher levels of engagement and creativity.
2. **Flexible work**: The second principle focuses on embracing flexibility in work arrangements. It acknowledges that traditional nine-to-five, location-bound work

models may not be suitable for the demands of the modern workforce. Flexible work arrangements, such as remote work or flexible hours, enable employees to achieve a better work-life balance, increase productivity, and reduce stress. Organizations are encouraged to provide the necessary support and infra-structure for flexible work practices.

3. **Collaboration and participation**: This principle emphasizes the value of collaboration and active participation within the organization. It promotes interdisciplinary teamwork, knowledge sharing, and open communication channels. By fostering a collaborative culture, organizations can harness the diverse perspectives and expertise of their employees, leading to innovation and better problem-solving.
4. **Lifelong learning**: The fourth principle highlights the importance of continuous learning and development. It recognizes that in a rapidly changing world, employees need to constantly ac-quire new skills and knowledge to stay relevant. Organizations are encouraged to provide opportunities for ongoing training, mentorship programs, and learning initiatives that support personal and professional growth.
5. **Shared value**: The final principle integrates the concept of shared value into the NWC. It emphasizes that organizations should actively seek ways to create value not only for themselves but also for society and the environment. By aligning business goals with social and environ-mental objectives, organizations can contribute to sustainable development, address societal challenges, and build a positive reputation.

These principles collectively provide a framework for organizations to implement New Work practices that promote employee well-being, flexibility, collaboration, learning, and social responsibility. By embracing these principles, organizations can create a modern and adaptive work culture that aligns with the demands of the cyber-physical reality.

Corporate Shared Value

New work organizations have realised that they are dependent from their environment; they follow the principles of Cybernetics (Beer, 1959, 1979, 1995). Therefore, new work organizations strengthen togetherness and their connection to the society and the local environment around them. This has to be performed via a sustainable way of pursuing business and increased efforts in the improvement of the region the company is located in. As Kamran (2021) pointed out, "Marketing today, however, needs to integrate for the well-being of the organization also the well-being of the societies". This integration is often misunderstood as a mandatory aspect that has to be fulfilled to be accepted by society and politics. The traditional view on corporate social responsibility lacks an underlying economic motivation for organizations to implement it. Therefore, the author proposes Porter and Kramers (2011) notion of shared value as the fifth element of a new work framework. It includes three central aspects that organizations should pursue to increase their sustainability while simultaneously increasing productivity and long-term profit generation. First, the organization should participate in the formation of local clusters. Those clusters include the improvement of the nearby region's infrastructure. If the educational environment is sophisticated, for example, the company has access to a pool of high professionals which simultaneously decreases training costs and increases productivity. A better transport infrastructure simplifies the logistic processes, and a general improvement of the regional poverty increases the health of the employees leading to less absence from work and again higher profits through increases productivity. Collaboration of firms does also play a central role in the building of clusters to combine competences and create synergies. An example for that is the local automotive cluster in southern Germany (Porter & Kramer, 2011). The next aspect, redefining productivity refers to the goal of uniting environmental and social sustainability with increasing productivity. Instead of exploiting suppliers in low-wage countries and making them more and more unproductive, companies can try to support their suppliers to increase their competences and efficiency. The resulting long-term effect of the increasing efficiency are products with a higher quality at lower prices which again embodies value creation through sustainable acts. Another example for that could be the creation of anti-smoking campaigns for employees to reduce health-related costs (Porter & Kramer, 2011). The last point refers to the modification of products. By manufacturing products that meet environmental and social necessities, companies can increase profits. This can be especially observed in the food sector which now put its focus on the health and nutrition of customers instead of quantity and test which perfectly suits the new trends in the food marketing sector, as pointed out by Kamran (2021). With such products, the overall societal well-being is improved by simultaneously finding niches to market the products (Porter & Kramer, 2011). If those aspects are kept in mind, organizations can effectively transform their work ethos to the dimension of New Work which is from essential importance for the implementation of new work principles. By incorporating the principles of cybernetics, new work organizations recognize their dependence on the environment and strive to strengthen their connections to society and the local community.

To achieve this, organizations focus on sustainable business practices and actively contribute to the improvement of the regions in which they operate. They participate in the formation of local clusters, collaborating with other firms to combine competences and create synergies. This collaboration leads to improved infrastructure, such as educational resources and transportation, which benefits both the organization and the community. The presence of a skilled workforce in the region reduces training costs and increases productivity, while better transportation infrastructure streamlines logistics. Additionally, efforts to alleviate regional poverty contribute to the well-being of employees, resulting in reduced absenteeism and increased productivity.

The concept of shared value reinforces the economic motivation for organizations to implement corporate social responsibility. It suggests three central aspects that organizations should pursue to enhance sustainability and profitability simultaneously:

1. **Participating in the formation of local clusters**: By actively engaging in local clusters, organizations can foster collaboration, share resources, and contribute to the development of the region's infrastructure. This benefits the organization by providing access to a skilled workforce and improving logistical efficiency.
2. **Redefining productivity**: Instead of exploiting suppliers in low-wage countries, organizations can support their suppliers to increase competence and efficiency. This leads to higher-quality products at lower costs, resulting in value creation through sustainable practices. For example, organizations can invest in anti-smoking campaigns to reduce health-related costs and improve employee well-being.
3. **Modifying products**: By manufacturing products that align with environmental and social needs, organizations can increase profitability and contribute to societal well-being. An example is the shift in the food sector toward healthier and more sustainable products, catering to the changing preferences of customers.

By integrating these aspects into their operations, organizations can effectively transform their work ethos and align with the principles of new work. This holistic approach not only benefits the organization but also creates a positive impact on society and the environment.

4 Cyber-Physical Navigator Model

After elucidating the diverse components and systems that facilitate organizational transformation to address the challenges posed by the cyberphysical reality and the complexities arising from the COVID-19 pandemic, the subsequent endeavor involves contextualizing these elements within the do-mains of cyberphysical and designerly marketing to establish an adequate control system. Coined as the Cyber-physical Networked Marketing (CNM), this control system accentuates its capacity to guide organizations through intricate and disruptive environments. A convergence of sophisticated Augmented Reality (AR), Virtual Reality (VR), and Artificial Intelligence (AI) technologies, together with top-tier management models grounded in fundamental sciences like cybernetics and design, is deemed in-dispensable for ensuring future organizational resilience. Echoing Ashby's axiom, "only variety can ab-sorb variety" (Ashby, 1956), the CNM incorporates a requisite level of variety through the integration of multiple frameworks within the organizational dimension, management unit, and management sys-tem to effectively address the multifarious perturbations arising from the cyberphysical environment (Kamran, 2019).

The CNM emerges as a potent control system designed to provide organizations with a comprehensive toolkit, synergizing advanced technologies such as AR, VR, and AI with effective management models. It judiciously recognizes the intricacies and disruptive nature of the cyberphysical reality, which have been further accentuated by the challenges posed by the COVID-19 pandemic. By drawing insights from cybernetics and design thinking, the CNM offers a coherent framework that empowers organizations to navigate this environment and adeptly respond to dynamic circumstances.

Central to the CNM's underpinning philosophy is the principle of "variety." Inspired by Ashby's law of requisite variety (Ashby, 1956), the CNM acknowledges the need for organizations to possess a requisite diversity in their control systems to effectively confront and assimilate the diverse challenges and perturbations encountered within the cyberphysical environment. To achieve this, the CNM seamlessly integrates multiple frameworks across various organizational dimensions, management units, and management systems. These frameworks furnish indispensable tools and strategies to enhance organizational resilience and adaptability. By harnessing the potential of advanced technologies such as AI, organizations can glean invaluable insights, optimize marketing strategies, and make data-driven decisions.

Moreover, the CNM places utmost emphasis on interdisciplinary collaboration and knowledge sharing. By fostering synergies among experts from diverse domains, including marketing, technology, and design, organizations can harness manifold perspectives to engender innovative solutions and stay ahead in the cyberphysical reality.

The AGMI model, which centers on AI and its implications for marketing, serves as a tangible illustration of how the CNM approach can be effectively implemented. By seamlessly integrating AI technologies into marketing practices, organizations can elevate customer experiences, personalize marketing endeavors, and elevate overall organizational performanceFig. 13.

In summation, the CNM presents an all-encompassing and integrated approach to control systems and organizational resilience within the domain of the cyberphysical reality. Leveraging advanced technologies, management models, and interdisciplinary collaboration, organizations can adeptly navigate the intricacies of the digital landscape and propel sustainable success into the future.

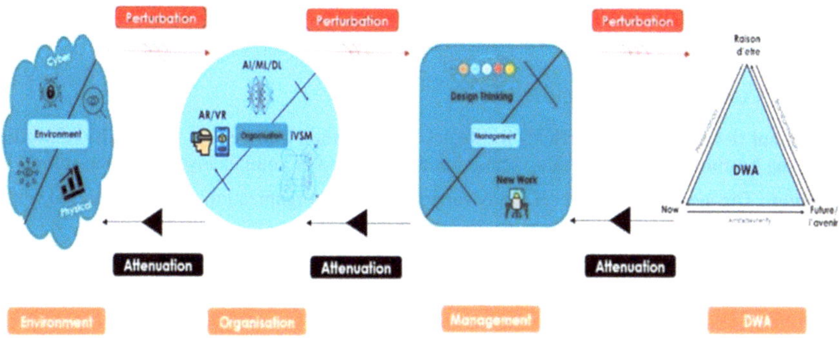

Fig. 13 Cyber-physical navigator model

5 Conclusion

The proposed ontological and managerial logic within the systems of the CNM is the DWA The DWA encompasses a designerly view that highlights the essential elements of preservation, ambidexterity, and transformation within the context of the cyber-physical environment. It serves as a complementary framework to cybernetics, viable systems, Design Thinking (DT), and new work, providing a comprehensive rationale for managerial decision-making.

The DWA recognizes the importance of striking a balance between preservation, which involves optimizing and maintaining existing systems and processes, and ambidexterity, which entails the ability to explore new opportunities and adapt to changing circumstances. Additionally, it emphasizes the significance of transformation, which encompasses ongoing innovation and the ability to reimagine and re-design organizational systems and practices.

By adopting a designerly way of acting, managers are encouraged to approach decision-making with a focus on creativity, user-centricity, and systemic thinking. The designerly view prompts managers to consider not only the current operational context but also the future trajectory and purpose of the organization. It provides a framework for envisioning and implementing innovative solutions and strategies.

Furthermore, the DWA facilitates the integration of the central unit of the proposed system, the VR/AR/AI machine interface, by offering convenience and support. By embracing a designerly approach, managers can gain a better understanding of user needs, design intuitive interfaces, and lever-age advanced technologies such as Virtual Reality (VR), Augmented Reality (AR), and Artificial Intelligence (AI) to enhance user experiences and drive organizational success.

In conclusion, the DWA presents a comprehensive and designerly perspective that aligns with the principles of the CNM. By adopting this approach, managers can effectively navigate the complexities of the cyber-physical environment, foster innovation, and ensure the seamless integration of advanced technologies into their organizational systems.

Authors' Contributions Not applicable.

Additional declarations for articles in life science journals that report the results of studies involving humans and/or animals. There were no humans or animals involved in the study.

Funding The authors did not receive any support from any organization for the submitted work.

Availability of Data and Material Not applicable.

Code Availability Not applicable.

Conflicts of Interest/Competing Interests The authors have no conflicts of interest to declare that are relevant to the content of this article.

Ethics Approval Not applicable.

Consent to Participate Not applicable.

Consent for Publication Not applicable.

References

Ahmaniemi, T., Lindholm, H., Muller, K., & Taipalus, T. (2017). Virtual reality experience as a stress recovery solution in workplace. In *IEEE Life Sciences Conference* (pp 206–209). https://www.researchgate.net/publication/322077358_Virtual_Reality_Experience_as_a_Stress_Recovery_Solution_in_Workplace

Alexander, C. (1964). *Notes on the synthesis of form*. Harvard University Press.

Archer, B. (1965). *Systematic method for designers*. The Design Council.

Archer, B. (1979). Design as a discipline. *Design Studies, 1*, 17–20. https://doi.org/10.1016/0142-694X(79)90023-1

Ashby, W. R. (1956). *An introduction to cybernetics*. Chapman & Hall Ltd.

Asimow, M. (1962). *Introduction to design*. Prentice-Hall.

Baker, G. (2020). *Virtual reality comes to Virbela*. Retrieved August 17, 2020, from https://www.virbela.com/blog/virtual-reality-comes-to-virbela

Barnes, S. (2016). Understanding virtual reality in marketing: Nature, implications and potential. *SSRN Journal*. https://doi.org/10.2139/ssrn.2909100

Beckman, S. L., & Barry, M. (2007). Innovation as a learning process: Embedding design thinking. *Cali-Fornia Management Review, 50*, 25–56. https://doi.org/10.2307/41166415

Beer, S. (1959). *Cybernetics and management*. English Universities Press.

Beer, S. (1979). *The heart of enterprise*. The Stafford Beer classic library. Wiley.

Beer, S. (1984). The viable system model: Its provenance, development, methodology and pathology. *The Journal of the Operational Research Society, 35*, 7–25. https://doi.org/10.1057/jors.1984.2

Beer, S. (1995). *Brain of the firm*. The managerial cybernetics of organization (2nd ed.). Wiley.

Bergmann, F. (1977). *On being free*. University of Notre Dame Press.

Bergmann, F. (2019). *New work, new culture: Work we want and a culture that strengthens us*. Zero Books.

Brenner, W., & Uebernickel, F. (2016). *Design thinking for innovation: Research and practice* (1st ed.). Springer International Publishing.

Brown, T. (2008). Design thinking. *Harvard Business Review*.

Carruth, D. W. (2017). Virtual reality for education and workforce training. In *2017 15th International Conference on Emerging eLearning Technologies and Applications (ICETA)* (pp. 1–6). IEEE.

Christopher, W. F. (2007). *Holistic management: Managing what matters for company success*. Wiley series in systems engineering and management. Wiley-Interscience.

Cross, N. (1982). Designerly ways of knowing. *Design Studies, 3*, 221–227. https://doi.org/10.1016/0142-694X(82)90040-0

Cross, N. (1993). A history of design methodology. In *Design methodology and relationships with science* (pp. 15–27).

Cross, N. (1999). Design research: A disciplined conversation. *Design Issues, 15*, 5. https://doi.org/10.2307/1511837

Dale, E. (1969). *Audio-visual methods in teaching* (3rd ed.).

Davies, W. (2020). *The last global crisis didn't change the world. But this one could*. Retrieved January 8, 2020, from https://www.theguardian.com/commentisfree/2020/mar/24/coronavirus-crisis-change-world-financial-global-capitalism

Frewen, P., Mistry, D., Zhu, J., Kielt, T., Wekerle, C., Lanius, R. A., & Jetly, R. (2020). Proof of concept of an eclectic, integrative therapeutic approach to mental health and well-being through virtual reality technology. *Frontiers in Psychology, 11*, 858. https://doi.org/10.3389/fpsyg.2020.00858

Hackl, B., Wagner, M., Attmer, L., Baumann, D., & Zünkeler, B. (2017). *New Work: auf dem Weg zur neuen Arbeitzwelt: Management-Impulse, Praxisbeispiele, Studien.* Springer Gabler, Wiesbaden.

Hall, A. D. (1962). *A methodology for systems engineering.* Van Nostrand.

Hassi, L., & Laakso, M. (2011). Conceptions of design thinking in the design and management discourses: Open questions and possible directions for research. In *Proceedings of IASDR2011.*

Helmold, M. (2021). *New Work, transformational and virtual leadership: Lessons from COVID-19 and other crises.* Springer International Publishing AG.

Hilken, T., Heller, J., Chylinski, M., Keeling, D. I., Mahr, D., & de Ruyter, K. (2018). Making omnichannel an augmented reality: The current and future state of the art. *Journal of Research in Interactive Marketing, 12*, 509–523. https://doi.org/10.1108/JRIM-01-2018-0023

Johansson-Sköldberg, U., Woodilla, J., & Çetinkaya, M. (2013). Design thinking: Past, present and possible futures. *Creativity and Innovation Management, 22*, 121–146. https://doi.org/10.1111/caim.12023

Kamran, Q. (2019). *Developing a holistic model for competitive strategic management.* Doctoral Thesis, University of Latvia.

Kamran Q (2021) *Managing complexity in marketing from a design Weltanschauung.* Doctoral thesis, University of Twente.

Malik, F. (2011). *Strategie: Navigieren in der Komplexität der Neuen Welt* (Management: Komplexität meistern). Management: Komplexität meistern (Vol. 3, 1st ed.). Campus.

Michalos, G., Karvouniari, A., Dimitropoulos, N., Togias, T., & Makris, S. (2018). Workplace analysis and design using virtual reality techniques. *CIRP Annals, 67*, 141–144. https://doi.org/10.1016/j.cirp.2018.04.120

Naylor, M., Morrison, B., Ridout, B., & Campbell, A. (2019). Augmented experiences: investigating the feasibility of virtual reality as part of a workplace wellbeing intervention. *Interacting with Computers, 31*, 507–523. https://doi.org/10.1093/iwc/iwz033

Ozturkcan, S. (2020). Service innovation: Using augmented reality in the IKEA Place app. *Journal of Information Technology Teaching Cases* 204388692094711. https://doi.org/10.1177/2043886920947110

Plattner, H., Meinel, C., & Weinberg, U. (2009). Design thinking: Innovation lernen - Ideenwelten öffnen. mi-Wirtschaftsbuch, FinanzBuch, München.

Porter, M. E. (1985). *Competitive advantage.* The Free Pess.

Porter, M.E., & Heppelmann, J. E. (2017). *Why every organization needs an augmented reality strategy.* Retrieved May 25, 2020, from https://hbr.org/2017/11/a-managers-guide-to-augmented-reality

Porter, M. E., & Kramer, M. R. (2011). Creating shared value: How to reinvent capitalism—And unleash a wave of innovation and growth. *Harvard Business Review* 1–17.

Simon, H. (1969). *The sciences of the artificial.* Cambridge, MA.

Sutherland, J., Belec, J., Sheikh, A., Chepelev, L., Althobaity, W., Chow, B. J. W., Mitsouras, D., Christensen, A., Rybicki, F. J., & La Russa, D. J. (2019). Applying modern virtual and augmented reality technologies to medical images and models. *Journal of Digital Imaging, 32*, 38–53. https://doi.org/10.1007/s10278-018-0122-7

Tschepe, S. (2017). *Was sind die wichtigsten Eigenschaften und Fähigkeiten von Design Thinking-Coaches?* Master thesis, Humboldt-Universität zu Berlin.

Tushman, M. L., & O'Reilly, C. A. (1996). Ambidextrous organizations: Managing evolutionary and revolutionary change. *California Management Review, 38*, 8–29. https://doi.org/10.2307/41165852

Väth, M. (2016). Arbeit: Die schönsten Nebensache der Welt: wie New Work unsere Arbeitswelt re-volutioniert. GABAL, Offenbach am Main.

Väth, M., Gstöttner, A., & Soballa, A. (2019). *New Work Charta*. Retrieved Februarry 8, 2021, from https://humanfy.de/new-work-charta/
Wiener, N. (1948). *Cybernetics: Or communication and control in the animal and the machine*. MIT Press.

Responsible Management: Structural Components of Business Ethics

Mine Afacan Findikli

Action comes_not from thought, but from a readiness for responsibility
D. Bonhoef
*German Lutheran pastor, theologian

Abstract Recent research highlights three key attributes shared by organizations prepared for the future. These attributes encompass clearly understanding their identity and values, prioritizing efficiency and agility, and enhancing their capacity for learning and innovation to facilitate growth. Additionally, nowadays effective management incorporates also sustainability, responsibility, and ethical considerations, alongside technological advancements, to promote responsible practices. These factors might be called in the name of sustainability management. This chapter's objective is to concentrate on the structural aspects of business ethics, aiming to cultivate responsible management.

Keywords Responsible management · HRM practices · Ethical leadership · Ethical culture

1 Introduction

Businesses consider the needs of society, both in terms of the economy and social aspects, as well as the environment, as they are one of the important building blocks of the society in which they operate. Therefore, serving the goals of sustainable development is an important dimension of their corporate strategies while conducting their economic activities. Previously, businesses that demonstrated their good corporate citizenship through social responsibility projects are now expected to not only showcase the outputs of their contributions but also internalize their values based on business ethics in their processes and lead the increase in the welfare of society.

M. A. Findikli (✉)
FEASS, Business Department, Istinye University, İstanbul, Turkey
e-mail: mine.findikli@istinye.edu.tr

C. Machado (ed.), *Building the Future with Human Resource Management*,
Management and Industrial Engineering, https://doi.org/10.1007/978-3-031-52811-8_5

Following the publication of the report titled "Our Common Future," also known as the Brundtland Report, by the World Commission on Environment and Development in 1987, businesses have been determining their corporate sustainability strategies in line with the social, environmental, and economic goals of Sustainable Development (Keeble, 1988). In this context, businesses, including global enterprises, announce that they plan their production and business processes in accordance with international and national variables, policies, and relevant legal regulations. However, in an increasingly uncertain, complex, and volatile business world, it is not easy for businesses to successfully fulfill their corporate social responsibilities and improve their financial performance. In parallel, if top management does not sustain their corporate and competitive strategies within the framework of ethical principles and values, their CSR practices within the scope of SDGs may suffer setbacks and their corporate reputation may be significantly damaged in the eyes of the society. In line with this thought, it is helpful to recall two examples from the past decade.

In the second week of September 2015, news outlets worldwide reported a "major scandal" involving the Volkswagen Group, known for using deceptive software, which came to be known as the emission cheating scandal. It was revealed that certain models of the Volkswagen Group had been equipped with manipulative software to deceive emission tests, resulting in lower recorded levels of environmental pollution. This manipulation aimed to deceive the U.S. Environmental Protection Agency. The irresponsible and utilitarian approach of the top management led to the initial resignation of CEO Martin Winterkorn. Furthermore, as a result of the fraudulent emission manipulation, the German company's stock experienced a decline during that period, and the company had to recall 11 million vehicles from the market (Hotten, 2015).

The unethical behavior employed by Volkswagen to gain a larger market share in the American market not only drew condemnation from the environmentally conscious society but also resulted in legal penalties. According to announcements by the US Department of Justice (2016), VW had to spend up to $10 billion to recall 2.0-L vehicles from the market, terminate leases, and compensate consumers for their losses. Additionally, the company made a $4.7 billion investment to support zero-emission vehicle technology. The statement by Deputy Attorney General S. Q. Yates (2016) from the US Department of Justice, highlighting that the company had made approximately half a million American drivers unwitting partners in an unprecedented attack on the environment by deceiving regulators, was crucial in emphasizing the importance of responsible management from a sustainability perspective.

When evaluating the scandal from a perspective of responsible management, it wouldn't be wrong to say that even some global businesses, driven by the objective of increasing profitability and expanding their market, disregard responsible management based on business ethics without considering ecological balance in a utilitarian approach, as seen in this example. Another example of how society can be manipulated is the Facebook-Cambridge Analytica Data Scandal of 2018, where ethical considerations were neglected, and the boundaries of responsible management were blurred and stretched. In the 2010s, approximately 87 million Facebook users' personal data was collected by Cambridge Analytica without their consent (Meredith, 2018). Cambridge Analytica was a political consulting firm that used data mining techniques to provide analytical support to the 2016 presidential campaigns of Ted Cruz and Donald Trump (Smith, 2023).

Although Facebook has announced principles and rules within the framework of ethics and responsible management in its "Code of Conduct and Facebook's Data Policy" (e.g., Code of Conduct- Sect. 8: Protection of User Data and Personnel Data, 2019), it violated this clause of the code by being aware of Cambridge Analytica's misuse of user data but failing to take any action to stop them. According to Forbes Magazine's article titled "The Problem Isn't Cambridge Analytica: It's Facebook," one of the most striking conclusions that can be drawn from the news surrounding the 2016 elections was the sharp shift in public and media reaction to the political use of data-driven election targeting since the 2008 and 2012 Obama campaigns (Leetaru, 2019). Even during the Obama campaign, DVRs were used to track what each voter watched on television. In short, according to the author of the Forbes article, in the Facebook-Cambridge Analytica Data Scandal, data was considered good, and pushing the boundaries of data privacy and ethics to monitor and manipulate voters was not seen as a negative behavior (Leetaru, 2019).

The concept of responsible management is a commitment by businesses to make positive and sustainable impacts by making business decisions that are aligned with business ethics and values. This approach aims to integrate sustainability, responsibility, and ethics into management practices (Gherardi & Laasch, 2022). In this section, we will discuss the structural components that help translate business ethics, which form the foundation of responsible management but often remain limited to rhetoric, into practices through business strategies and implementation.

2 Responsible Management

Milton Friedman's perspective, which argues that the sole purpose of businesses is to maximize returns for their shareholders, has been replaced by Carroll's (1991) model of corporate social responsibility (CSR) and the "Triple Bottom Line" (TBL) concept introduced by John Elkington in 1999, which has been integrated into literature and business practices for over three decades. Carroll's corporate social responsibility pyramid consists of four successive levels of responsibility: economic, legal, ethical, and philanthropic (Carroll, 1991). On the other hand, Elkington and Rowlands (1999) perspective of the "Triple Bottom Line" is developed based on the idea that businesses should focus not only on creating financial value but also on preserving social and ecological balance. It is a management concept that advocates for businesses to commit to measuring their social and environmental impacts in addition to their financial performance. The concept encompasses the "Three Ps": profit, people, and planet (Preira & Martins, 2021).

It can be said that Carroll's (1991) concept of corporate social responsibility and Elkington and Rowlands (1999) "Triple Bottom Line" concept lay the foundations of the responsible management perspective. Naturally, in a world where profitability increase is often considered more important than the purpose itself within a capitalist system, businesses developing their corporate strategies based on business ethics and responsibility, with a focus on sustainable development goals, constitute the scope of responsible management that reflects in business practices (Gherardi & Laasch, 2022).

Business ethics is concerned with the ethical correctness or incorrectness of decisions and behaviors in the context of work and management processes. Within this framework, business ethics relates to (1) employees' decision-making and behaviors regarding work processes and practices, (2) the organization's decisions and policies concerning external environment and stakeholders, and (3) the ethicality of the policies and strategies exhibited by the business within the competitive market and the business ecosystem (Afacan Findikli, 2017). Particularly, after cases such as Enron, WorldCom, and Tyco International, sensitivity towards business ethics has increased worldwide (Selvarajan, 2006). Since the 1980s, many businesses, initially in the United States, have started establishing "ethical committees" and announcing their "ethical codes." As a result, companies have come under scrutiny not only for their financial performance but also for their ethical behavior in light of evolving societal awareness and legal obligations. In this regard, business ethics aligns with corporate social responsibility and the two are interconnected. Business ethics involves utilizing ethical principles and values in the decision-making process to determine the correct and incorrect path, while corporate social responsibility encompasses the ethical behavior and corporate citizenship that a business develops towards all its stakeholders (Carroll, 1991).

The element that makes business ethics and corporate social responsibility so similar is values (Rushton, 2002). The resemblance between the two concepts is also evident in Carroll's model of the corporate social responsibility pyramid (CSR). The pyramid consists of four successive levels of responsibility. As the levels ascend, the awareness of responsibility increases, encompassing the general public. These levels are economic, legal, ethical, and philanthropic responsibilities (Carroll, 1991). The economic responsibility mentioned by Carroll corresponds to the "stakeholder theory" proposed by Milton Friedman, where the considerations are given to shareholders and business owners. Friedman established a "social contract" between businesses and their stakeholders (Friedman, 1962). According to the contract, businesses provide benefits to shareholders and top management through economic development. Economic development equates to profitability. However, in the modern approach, the social contract, which is one of the foundations of sustainability, argues that businesses should have social obligations towards society beyond profitability. These social obligations towards society aim to enhance the welfare of the community both economically and socially, and also include practices to prevent ecological issues. The third level, ethical responsibilities, encompasses economic and legal responsibilities and goes beyond them, focusing on the company's provision of proper, good, and acceptable policies and strategies to society. The company's compliance with regulations and rules, as well as fulfilling its responsibilities towards employees, society, and the environment in accordance with the legal framework of the community it operates in, are part of this level. In the fourth level, the business internalizes corporate citizenship behaviors and develops strategies and policies that are beneficial to society and the environment while fulfilling its commercial activities in a voluntary manner (Carroll, 1991).

The field of sustainability extends to the concept of sustainable development, which is defined as "meeting the needs of the present without compromising the ability of future generations to meet their own needs" (UN, 1987). In today's world, the aspects of sustainability that businesses need to pay attention to while meeting their own needs were outlined at the 2005 United Nations World Summit. These aspects include, as mentioned in previous sections, economic development, social development, and environmental protection (UN, 2005). These goals are not only the responsibilities of governments and official institutions but also of businesses, which are important components of the system. According to an article in Forbes titled "Setting The Right Sustainability Goals" (Steele, 2019), 181 CEOs of global businesses came together to collectively evaluate the steps that can be taken for sustainability. The main theme of the meeting was to aim for sustainability in the twenty-first century by creating shared and long-term value for the common welfare of societies. This example highlights the importance of taking collective action and practicing responsible management in response to the negative impacts of economic, social, and environmental crises.

It is clearly known that it is necessary for businesses to progress with a responsible management approach within the framework of the market conditions in which they compete. However, the adoption of business ethics and responsibility as fundamental to sustainability goals is challenged by the pressure of rapidly changing external environmental dynamics. Nevertheless, significant steps have been taken, particularly by global companies. In recent years, it can be observed that businesses serve values shaped by their ethical codes within the framework of the Sustainable Development Goals (SDGs). For example, Coca-Cola demonstrated its sensitivity to gender equality within the community by signing the "Empowerher" declaration in April 2023. Vodafone aims to combine the ideas, production, technology, and commercial knowledge of small entrepreneurs and social entrepreneurs through its "Business Ventures: social enterprise" structure to improve lives. The sharing of Vodafone Business Ventures (Vodafone, 2015) on corporate websites is highly valuable in terms of responsible management. *Vodafone is proud to have made a positive impact on people's lives for the betterment of society, and they are fully committed to continuing this work. They recognize that customers now expect them to act ethically, as failing to do so may result in a loss of loyalty.*

In light of the given examples, the importance of institutionalizing business ethics and internalizing them by stakeholders in responsible management for serving the goals of sustainable development once again becomes evident.

3 Structuring Business Ethics Within Responsible Management

In today's world, it is not sufficient for businesses to be considered responsible solely based on the ethical practices within their own boundaries. It is essential that all business partners involved in the value chain also conduct their commercial transactions with similar ethical principles and values. In this regard, businesses are not just profit-driven entities; they are accountable moral entities to their employees, investors, suppliers, and customers. They are responsible for the decisions they make, the consequences of those decisions, and the behavior of their employees. From this perspective, the Ethics Leadership Institute defines ethical business as "a community of individuals working together in an environment based on mutual respect, where everyone can develop personally, contribute to common interests, and share the personal, emotional, and financial rewards of a job well done" (Marcouse, 2014). However, as can be understood from a few negative examples provided above, simply having established ethical principles and values is not sufficient for businesses to shape their activities and employee behavior with sustainability and responsibility in mind. Responsible Management requires the adoption of a leadership style that aligns with this understanding, the internalization of ethical codes through human resource management practices, the cultivation of an organizational climate nourished by a responsible management approach, and the planning of supply chain systems.

3.1 Ethical Leadership in the Context of Responsible Management

According to upper echelons theory, the top management team (TMT) plays a key role in leading the process of formulating and implementing corporate strategy (Lu et al., 2022). Based on this, the question arises as to how leadership attitudes and behaviors motivate businesses towards sustainable development goals in line with a responsible management approach, transform organizational culture, and shape the behavior of followers. The answer to this question can be addressed within the scope of green transformative leadership (Begum et al., 2021), transformative leadership (Pham et al., 2022), and environmental leadership (Zhang & Ma, 2021). Additionally, it would be appropriate to discuss leadership behavior within the framework of ethical leadership (Constantinescu & Kaptein, 2021), drawing from both Aristotelian virtue ethics based on philosophical virtue development and the positive organizational behavior framework.

Management practices designed with a focus on social development and ecological business ethics not only extend to the practices within a business's own boundaries but also positively impact the practices and relationships of all stakeholders within the entire supply chain (Jamali, 2008). However, for responsible management policies to be effective in achieving sustainable development goals in the long term, corporate leadership needs to transform organizational culture based on values, mission, and vision. Leadership, recognized as one of the most important resources for businesses within the resource-based theory (Barney, 2001), is a key factor in the transformation towards responsible management. In this context, an ethical leader can effectively lead the connection of environmental projects for environmental health and safety and relevant projects for increasing social welfare with corporate strategies, serving responsible management. Throughout this process, a set of values that will provide guidance to employees within the business can be identified and disseminated.

Ethical leadership is often described in the literature in terms of charismatic and transformative characteristics. Transformative leaders align their own and their followers' value systems with ethical principles (Burns, 1978), serve as role models in upholding high standards of ethical and moral behavior, and influence their employees to change their behavior (Avolio, 1999). Brown and Trevino (2006) define an ethical leader as someone who normatively behaves appropriately in mutual interactions and guides their followers towards such behavior through these interactions. This definition encompasses both ethical conduct and being a role model, making it a leadership style that can serve the principles of responsible management. Ethical leadership, which emerged with the intention of eliminating corporate misconduct (Kuenzi et al., 2020), is strategically important due to its positive effects on subordinates and the organizational benefits it generates (Fontrodona & Berrone, 2018). Ethical leaders building high-quality relationships with their subordinates, being trustworthy and legitimate role models, creating a fair working environment, and actively managing ethics in the workplace serve as significant sources of motivation for their followers to develop normative behaviors (Brown & Treviño, 2006). The responsible management approach demonstrated by ethical leaders can enhance employee creativity and foster prosocial behavior towards society, thus playing a crucial lever in achieving sustainable development goals (Liu & Zhao, 2019). Over the past decade, the impact of ethical leadership and ethical culture on environmental innovation capacity (Van Der Wal & Demircioğlu, 2020; Liu & Zhao, 2019) and green innovation behaviors has been confirmed (Wood et al., 2021; Şengüllendi et al., 2023).

On the other hand, Green transformational leadership has been identified as the most effective leadership style where the leader seeks to change the status quo by transforming the organization and also considers achieving environmental goals and inspiring followers to perform beyond expected levels of environmental performance (Ayandibu, 2019). "Green transformational leadership" is defined as "behaviors of leaders who motivate followers to achieve environmental goals and inspire followers to perform beyond expected levels of environmental performance" (Chen & Chang, 2013; Chen et al., 2014). Managers who assume a bridging role between top management and employees influence the behavior of their subordinates through their leadership styles (Liu & Zhao, 2019). Similarly, transformative leadership and environmental leadership positively impact green organizational behavior and green innovation (Chen & Chang, 2013). Ethical leaders also have a similar impact by embodying ethical behavior and serving as role models for their employees. Employees learn by observing their leaders' behavior and acquire expected attitudes through mutual interactions within the framework of social learning theory (Bandura, 1977).

An example from the industry can be given with Cadbury Schweppes. The company, as the world's third-largest beverage producer, continues its financial success with its ethical behavior. The main focus of the company's social responsibility behavior is to improve the nutritional quality of products consumed by children. Additionally, managing diversity and differences in the workplace is a fundamental principle. The company diligently monitors efficient energy and production practices for environmental protection, including greenhouse gas emissions and wastewater treatment. In its relationships with suppliers, the company develops assistance programs for local communities and farmers from whom it sources raw materials. Cadbury Schweppes' business approach is based on ethical behavior and close relationships with stakeholder groups. The success of the company is attributed to meeting the expectations of its employees, customers, and other stakeholders and a strong commitment to ethical principles (Kumar & Singh, 2013). Relevant literature also confirms the positive impact of ethical leadership through sustainable ethical climate on performance (Dey et al., 2022).

Considering previous studies, it is understood that ethical leadership, by investing in social capital, primarily enhances trust, loyalty, and a sense of belonging among employees, improves their performance and self-efficacy, reduces unethical behavior, and supports an increase in positive organizational behaviors (Shafique et al., 2020). In today's world, ethical leaders can effectively communicate the organization's social responsibility motivation and practices to their followers. They can serve as a proper role model for responsibility, transparency, and accountability (Walumbwa et al., 2011; Choi et al., 2015) and also play a transformative role in this process. Previous studies have confirmed that ethical leadership positively influences voluntary responsible behavior among employees through ethical climate (Dey et al., 2022; Shin et al., 2015). The findings of a study conducted by Shin et al. (2015) with 4,468 employees from 147 Korean companies across various sectors indicated that top management's ethical leadership significantly influenced the ethical climate and fully mediated its effects on organizational citizenship behavior and the financial performance of the company at the organizational level. Similar findings have been observed in Chun et al.'s (2013) study, which highlights the positive impact of corporate ethics on a company's financial performance.

Top management and leadership not only shape ethical culture but ethical culture also shapes responsible behavior and guides employees' daily actions (Van der Wal & Demircioglu, 2020).

3.2 *Ethical Organizational Culture and Climate Supporting Responsible Management*

With the understanding of preserving global resources for future generations, global environmentalism has become a priority issue in society over the past two decades. Consequently, environmental management, being an environmentally friendly business and considering environmental health in production, has gained importance for businesses. On the other hand, improving social well-being, eliminating inequalities, and supporting women's participation in the economy are also within the scope of responsible management. Therefore, businesses guide their employees on what workplace behaviors align with sustainability, how to be motivated in this regard, and how to treat stakeholders. In this context, organizational culture provides information on "how things are done" in terms of values, principles, and norms that influence the behaviors of its members and shape employee behavior. Another concept often used in conjunction with organizational culture is organizational climate. The definition of organizational climate was first used by Argyris (1958) to describe the psychological structure of an organization. In this framework, organizational climate is the sum of the continuous characteristics of the internal environment that individuals can perceive through their senses and that influence the behavior and motivation of employees, distinguishing one organization from others (Litwin & stringer, 1968) . Ethical organizational culture and climate contribute to the continuation of activities with a sense of social responsibility, ensuring the legality of managerial decisions, gaining the trust of the community, and delivering services and products of desired quality standards. In this regard, a strong organizational culture and climate have an impact on the social performance of businesses. In an organization with strong ethical values, four common characteristics stand out: (1) Interacting with all internal and external stakeholders, (2) Acting fairly in all business processes and towards society, (3) Conducting all practices with a high sense of responsibility, (4) Engaging in activities with individuals and members who uphold high ethical values and staying aligned with objectives (Küçükoğlu, 2012).

Just like the British fashion brand Ted Baker... One of the key reasons for the business to be ahead of its competitors and have brand recognition is its commitment to sustainability as an ethical company. Firstly, the business ensured that its mission statements are not just phrases on the website but are internalized by all employees, fostering the adoption of the company's values. Additionally, employees were expected to adhere to high standards aligned with ethical principles and the company made efforts to work with individuals who are compatible with its ethical principles. In this context, the company established a "Conscience Team" focused on addressing social, ecological, and ethical issues internally, promoting developments in employee attitudes and behaviors that increase societal and environmental awareness. Furthermore, the business is a member of Made-by, a European non-profit organization that helps scrutinize ethical behavior in the production of all used products and conducted transactions. As Bartels et al. expressed, businesses can better manage and improve organizational ethics through an ethical climate. In the case of Ted Baker, the ethical culture and climate played a supportive role in enhancing employees' moral sensitivity and behaviors (Bartels et al., 1998).

- W. L. Gore and Associates, like W.L. Gore & Associates, has centered its business structure around the perspective of sustainability and conveyed its corporate culture to its employees and stakeholders through the values it has defined within this framework. As a result, employees within the organization have been redefined as "associates" who come together to achieve shared objectives. An environment has been created within work teams where each individual voluntarily contributes to the development of new products and the improvement of team performance. Consequently, the company has gained a high level of creativity and adaptability. The principles and values that have positively influenced the organization's structure, culture, and climate are summarized as follows: (1) Treating all stakeholders, including employees, suppliers, and customers, fairly, (2) Supporting and encouraging employees (business partners) in acquiring and developing knowledge, (3) Making and keeping promises, (4) Believing in the necessity of negotiating the issue/problem with others before making high-risk decisions.

The effective use of limited resources for a sustainable environment requires employees to regulate their behaviors with this awareness, and thus, a green organizational culture is considered an abstract resource that serves this purpose (Banerjee, 2002; Yang et al., 2017). Green organizational culture helps to transmit values and assumptions that support sustainable economic and ecological development within the organization, integrating ecological understanding into the entire functioning of the business and across all levels of employees (Wang, 2019; García-Machado & Martínez-Ávila, 2019). As a result, it can better integrate sustainability efforts into its employees and business processes and transform employee behavior (employee green behavior) (Norton et al., 2015; Van Der Wal & Demircioğlu, 2020). Ethical leadership indirectly influences the ethical beliefs and behaviors of both directly associated employees and other followers in indirect relationships through the social interaction process and shared understandings and values representing ethical culture elements.

In summary, the formal structure of organizational culture, which includes ethical policies (implemented ethical rules, authority structures, reward systems, and ethical training), and the informal structure consisting of language, stories, peer behavior, and ethical norms intertwine (Schein, 2010), resulting in increased shared and internalized responsibility and awareness among employees. With the support of an ethical organizational culture and perceived climate, employees gain the motivation needed to be more responsible, transparent, and achieve higher moral performance, activating their cognitive resources (Schaubroeck et al., 2012).

3.3 Responsible Management-Focused HRM Practices (SRHRM)

In the process of responsible management, ethical leadership is not naturally alone, and HRM professionals are the most important supporters in internalizing the concept of responsible management in the organization (Gond et al., 2012). In today's society, where the sensitivity towards sustainable development goals has increased, businesses have begun to focus on structuring socially responsible human resource management policies and practices as they strive to become more responsible corporate citizens through corporate social responsibility (CSR) projects (Barrena-Martínez et al., 2019). In this context, it can be observed that the scope of the CSR concept has expanded as the transition from corporate social responsibility to responsible management has taken place. While corporate social responsibility previously referred only to external responsibilities, and some businesses even viewed these projects as part of their corporate marketing strategies (Hadjikhani et al., 2016; Kleyn et al., 2012; Lauring & Thomsen, 2009), the concept of responsible management includes an internal responsibility dimension in its policies and practices. In fact, without this dimension, the responsibility of companies remains incomplete. According to the new understanding of CSR, the responsibility of companies should encompass all stakeholders. Based on the principle of ethical responsibility, businesses should not only seek support from certain external stakeholders, but they should also be able to meet the demands of their internal stakeholders. HR policies are at the core of current CSR policies and practices (De Stefano et al., 2018).

The inclusion of HRM practices within the scope of corporate social responsibility is not only aimed at creating national and international social, environmental, and economic value but also focuses on addressing the demands and needs of various stakeholders involved in business activities (Barrena-Martínez et al., 2019). HRM practices that focus on responsible management (SRHRM) aim to proactively enhance employees' work experiences and meet their personal and social expectations in ethical and socially responsible ways (Omidi & Dal Zotto, 2022).

Ethical leaders significantly influence their employees' sense of responsibility and awareness through their attitudes and behaviors (Brown & Treviño, 2006), while SRHRM policies and practices empower employees and direct them towards more socially responsible behaviors by increasing their dedication to corporate social responsibility activities (Powell et al., 2013; Lu et al., 2022). When the strong focus of top management and stakeholders on responsible management philosophy aligns with ethical leadership, the implementation of CSR policies and practices triggers behavioral changes among employees in line with responsible management strategies. Creating a trusting environment among employees and fostering a management approach that is responsive to the organization's issues, transparent, and rewards virtuous behavior are essential. As a result, the combination of ethical culture and the psychological contract developed by employees regarding this issue increases their dedication to sustainability-related areas (Carlson & Perrewe, 1995).

Additionally, HR professionals develop appropriate policies and practices to ensure that employees adhere to specific values and ethical behaviors in line with the company's policies. They maintain transparency in the organization's internal and external actions and promote an inclusive recruitment strategy that aligns with the company's values and social responsibility principles. They also foster employee commitment to environmental sustainability, social justice, and equality, and support them in generating ethical, responsible, and sustainable solutions. HR professionals play a vital role in creating a corporate culture and climate that enables employees to uphold the company's values and social responsibility principles (Obrad & Gherheş, 2022).

When examining studies related to SRHRM, it is evident that HRM practices are predominantly adapted from traditional HR functions (such as recruitment and selection, training and development, working conditions, performance evaluation, and rewards) and enriched with fair, inclusive, and supportive practices. In recruitment, examples of practices include the application of fair and ethical criteria, equal opportunities for workforce diversities (e.g., sexual minorities), and giving priority to local vulnerable individuals. In addition to various practices such as flexible working conditions, effective internal communication, equal opportunities, and fair evaluation processes, providing remote working conditions for disabled individuals with an inclusive approach can also be considered as examples of SRHRM. It is argued that all these practices, starting within the boundaries of the organization, contribute to the increase in societal well-being and the establishment of a sustainable, responsible, and ethical society (Omidi & Dal Zotto, 2022).

For example, Green HRM practices enhance employees' sense of environmental responsibility. GHRM, through functions such as green recruitment, green training, green performance, and green compensation, enables employees to feel more responsible towards the environment. By effectively managing green performance and compensation, and establishing standards, it can increase employees' awareness of corporate sustainability norms. Thanks to the positive feedback from SRHRM practices, employees develop as individuals who actively meet environmental requirements (Lu et al., 2022).

The research conducted by SHRM/Ethics Resource Center (2008) reveals that human resource policies and practices are the most significant resources for ethical behavior in organizations. Human resource practices include: (1) ensuring the internalization of ethical principles and codes through training, (2) continuously communicating the importance of business ethics and proper conduct through communication channels, (3) serving as role models for ethical behavior, and (4) teaching everyone the responsibility to take action when ethical violations occur, making them vital supporters of organizational strategies. Ethical training programs and initiatives should encompass all employees and managers. Human resource policies and practices support the ethical culture and climate within organizations by promoting transparency, fairness, honesty, and constructive behavior (SHRM/Ethics Resource Center, 2008, pp. 2–3). For instance, Caterpillar, a major producer of mining and construction equipment, collaborated with consulting firms to provide ethical training for its 95,000 employees. These training programs have been recognized as a fundamental requirement for ongoing employment. Caterpillar has shared its ethical principles with employees since 1974, and these values include excellence, teamwork, commitment, integrity, and sustainability (Fastow, 2014).

In summary, human resource managers play important roles in ensuring the sustainability of an organization's policies: (1) Strategic Partnership: Human resource managers act as strategic partners to the top management, not only in achieving operational efficiency and profitability but also in developing employees to fulfill responsibilities towards society. They approach the fulfillment of missions towards the community and the environment by supporting and balancing the interests of stakeholders and employees. (2) Inclusiveness: They take on a guiding role with an inclusive approach that encompasses all internal and external stakeholders, especially in improving and revitalizing the social and natural environment significantly affected by business operations (Lockwood, 2009).

3.4 Sustainable Supply Chain Management in Responsible Management

In today's consumer landscape, consumers make purchasing decisions not only based on the brand but also taking into consideration the supply chains that produce and deliver the products. Since approximately 50% of the value of a product is realized by suppliers, businesses cannot achieve their sustainability goals without involving their business partners, or in other words, their suppliers. In doing so, they evaluate not only their own activities but also the commercial activities of their suppliers in terms of economic, social, and environmental dimensions. In the literature, the implementation of supply chain management from a sustainability perspective is referred to as "Sustainable Supply Chain Management (SSCM)". The concept of sustainability, in some businesses, may be addressed solely in terms of its environmental dimension (Charter et al., 2001). Therefore, businesses primarily focus on: (1) Ensuring a high level of environmental efficiency and conservation of natural resources, (2) Emphasizing the selection of environmentally friendly transportation systems from sourcing to the delivery of products and services to end consumers, (3) Reducing the diversity of non-recyclable waste and incorporating recyclable materials into product design (Altuntaş & Türker, 2012). However, social sustainability involves meeting the expectations of society with the products and services offered in the market, economic sustainability entails reorganizing all business processes to reduce costs and ensure regular monitoring, and environmental sustainability relates to the development of environmentally sensitive policies and strategies with the awareness that resources are limited. The implementation of these three elements, as defined by the United Nations, is possible through focusing on economic development, social development, and integrating ecological ethics into strategies and policies. With this understanding, the philosophy of responsible management promotes the effective governance of the environmental, social, and economic impacts throughout the lifecycle of products and services through good governance practices (UN Global Compact, 2010).

It is important for businesses to effectively communicate their ethical values, commitment to sustainable development goals, and responsible management approach to their business partners within the supply chain. Evaluating and monitoring activities not only based on financial indicators related to the production and market introduction of products, but also in terms of sustainable behaviors, is necessary. The reconfiguration of supply networks in terms of sustainability serves to create long-term value for both the ecological and social environments through the products and services offered to the market and to protect the value created. For instance, businesses are expanding production and operational boundaries throughout their supply chains to reduce emissions and waste in response to increasing pressures from regulatory authorities, media, and society, all of which contribute to the preservation of ecological balance (Wang & Feng, 2022).

When examining the literature on the establishment of Responsible Management-oriented Sustainable Supply Networks, it is observed that creating a culture of responsibility within the network is one of the key elements in generating value for sustainability (Chorn et al., 2010). In this context, ethical leaders can shape the supply chain with a responsible management philosophy by demonstrating behavior aligned with the universal values of business ethics and sustainable development goals. They can promote responsible, transparent, and fair conduct in business practices while creating an atmosphere of mutual trust. Building upon the study "Practicing management wisely" by Hühn et al. (2020), ethical leaders can serve as key drivers in creating a collaborative community that facilitates serving society in line with sustainable development goals.

Wang and Feng's (2022) study provides an example that demonstrates how ethical leadership displayed by top management can lead to the transformation of the activities of supply chain stakeholders into environmentally friendly practices. The research findings indicate that Supply Chain Ethical Leadership (SCEL) significantly supports strategic integration, information integration, and operational integration in the process of green supply chain transformation. This finding is consistent with previous studies emphasizing the importance of ethical leadership in interorganizational relationship management. Ethical leadership plays a crucial role in integrating supply chain partners into green initiatives, thereby fostering a more responsible management approach among green supply chain partners. In conclusion, if we consider Thomas Jefferson's quote "The price of freedom is eternal vigilance" in terms of creating and sustaining shared value that forms organizational culture, we can say that establishing a culture of responsible management is not an organic process but a carefully planned management process.

4 Conclusion

The financial crisis that shook the first decade of the new millennium and numerous business scandals have highlighted the lack of ethical integrity in business management, with examples of short-term profitability being pursued. The negative examples in the business world have increasingly emphasized the value of business ethics. As a result, businesses have started implementing strategies and policies towards sustainability and have recognized the need to act responsibly not only towards their shareholders but also towards all stakeholders involved. This has brought responsible management to the forefront.

When considering the scope of responsible management, it is essential to develop strategies and policies for sustainable development goals. These encompass practices related to economic development, social development, and environmental protection. Accordingly, companies are paying attention not only to economic development but also to social development and environmental protection in their operations. However, despite this, there are still businesses in the market that have not reached this level of consciousness and solely focus on their own interests and maximizing benefits. In the business world, there are still examples of misconduct such as scandals related to accounting records and financial reporting, and deceptive use of software in emission tests to ignore environmental pollution in commercial activities. Fundamentally, in the business world, when it comes to personal gain and egoistic values, legal regulations remain limited. At this stage, the concept of responsible management, which focuses on sustainable development goals and combines with business ethics, can generate effective solutions for a more responsible, fair, and sustainable environment through management practices. When the mechanisms inherent in the capitalist system come into play, managers may face challenges and dilemmas in simultaneously achieving sustainable performance and increasing competitive advantage. One of the most effective tools to overcome these challenges is the establishment of corporate codes of conduct framed by ethical values and norms, along with ethical leadership that leads the way in creating an ethical culture and climate within the organization.

References

Afacan Findikli, M. (2017). Sürdürülebilirlik Perspektifinde İş Etiği. In I. Pekdemir (Ed.), *İşletmelerde Sürdürülebilirlik Dinamikleri* (pp. 362–391). The First Press.

Altuntaş, C., & Türker, D. (2012). Sürdürülebilir Tedarik Zincirleri: Sürdürülebilirlik Raporlarının İçerik Analizi. *Dokuz Eylül Sosyal Bilimler Enstitüsü Dergisi, 14*(3), 45–46.

Argyris, C. (1958). Some problems in conceptualizing organizational climate: A case study of a bank. *Administrative Science Quarterly, 2*(4), 501–520. https://doi.org/10.2307/2390797

Avolio, B. J. (1999). *Full leadership development: Building the vital forces in organizations*. Sage.

Ayandibu, A. O. (2019). Green transformational leadership and organizational behavior effectiveness. In *Contemporary multicultural orientations and practices for global leadership, in green transformational leadership and organizational behavior effectiveness* (pp. 225–244). IGI Global. https://doi.org/10.4018/978-1-5225-6286-3.ch012

Bandura, A. (1977). *Social learning theory*. Prentice Hall.

Banerjee, S. B. (2002). Corporate environmentalism: The construct and its measurement. *Journal of Business Research, 55*(3), 177–191. https://doi.org/10.1016/S0148-2963(00)00135-1

Barney, J. B. (2001). Resource-based theories of competitive advantage: A ten-year retrospective on the resource-based view. *Journal of Management, 27*(6), 643–650. https://doi.org/10.1177/014920630102700602

Barrena-Martínez, J., López-Fernández, M., & Romero-Fernández, P. M. (2019). Towards a configuration of socially responsible human resource management policies and practices: Findings from an academic consensus. *The International Journal of Human Resource Management, 30*(17), 2544–2580. https://doi.org/10.1080/09585192.2017.1332669

Bartels, K. K., Harrıck, E., Martell, K., & Strıckland, D. (1998). The relationship between ethical climate and ethical problems within human resource management. *Journal of Business Ethics, 17*(7), 799–804.

Begum, S., Ashfaq, M., Xia, E., & Awan, U. (2021). Does green transformational leadership lead to green innovation? The role of green thinking and creative process engagement. *Business Strategy and the Environment, 31*(1), 580–597. https://doi.org/10.1002/bse.2911

Brown, M. E., & Trevino, L. K. (2006). Ethical leadership. A review and future directions. *The Leadership Quarterly, 17*, 595–616. https://doi.org/10.1016/j.leaqua.2006.10.004

Burns, J. M. (1978). *Leadership*. Harper Row.

Carlson, D. S., & Perrewe, P. L. (1995). Institutionalization of organizational ethics through transformational leadership. *Journal of Business Ethics, 14*, 829–838.

Carroll, A. B. (1991). The pyramid of corporate social responsibility: Toward the moral management of organizational stakeholder. *Business Horizon, 34*(4), 39–48.

Charter, M., Kielkiewicz-Young, A., Young, A., & Hughes, A. (2001). *Supply chain strategy and evaluation*. Austrália: The Sigma Project.

Chen, Y. S., & Chang, C. H. (2013). The determinants of green product development performance: Green dynamic capabilities, green transformational leadership, and green creativity. *Journal of Business Ethics, 116*(1), 107–119. https://doi.org/10.1007/s10551-012-1452-x

Chen, Y. S., Chang, C. H., & Lin, Y. H. (2014). Green transformational leadership and green performance: The mediation effects of green mindfulness and green self-efficacy. *Sustainability, 6*(10), 6604–6621. https://doi.org/10.3390/su6106604

Choi, S. B., Ullah, S. M., & Kwak, W. J. (2015). Ethical leadership and followers' attitudes toward corporate social responsibility: The role of perceived ethical work climate. *Social Behavior and Personality: An International Journal, 43*(3), 353–365. https://doi.org/10.2224/sbp.2015.43.3.353

Chorn, B., Sisco, C., & Pruzan-Jorgensen, P. D. (2010). *The business case for supply chain sustainability a brief for business leaders* (p. 5). BSR.

Chun, J. S., Shin, Y., Choi, J. N., & Soo, K. M. (2013). How does corporate ethics contribute to firm financial performance? The mediating role of collective organizational commitment and organizational citizenship behaviour. *Journal of Management, 39*(4), 853–877. https://doi.org/10.1177/0149206311419662

Constantinescu, M., & Kaptein, M. (2021). Virtue and virtuousness in organizations: Guidelines for ascribing individual and organizational moral responsibility. *Business Ethics, the Environment and Responsibility, 30*(4), 801–817. https://doi.org/10.1111/beer.12373

De Stefano, F., Bagdadli, S., & Camuffo, A. (2018). The HR role in corporate social responsibility and sustainability: A boundary-shifting literature review. *Human Resource Management, 57*(2), 549–566. https://doi.org/10.1002/hrm.21870

Dey, M., Bhattacharjee, S., Mahmood, M., Uddin, M. A., & Biswas, S. R. (2022). Ethical leadership for better sustainable performance: Role of employee values, behavior and ethical climate. *Journal of Cleaner Production, 337*. https://doi.org/10.1016/j.jclepro.2022.130527

Elkington, J., & Rowlands, I. H. (1999). Cannibals with forks: The triple bottom line of 21st century business. *Alternatives Journal, 25*(4), 42. https://doi.org/10.1002/tqem.3310080106

Fastow, A. (2014). Do not let yourself be involved in a fraudulent business, play by the rules. In Marcouse et al. (Ed.), *The business book* (pp. 122–124). Dorling Kindersley Limited. ISBN: 978-1-4654-1585-1

Fontrodona, J. R., & Berrone, P. (2018). Ethical challenges in strategic management: The 19th IESE international symposium on ethics, business and society. *Journal of Business Ethics, 152*, 887–898.

Friedman, M. (1962). *Capitalism and freedom*. The University of Chicago Press.

García-Machado, J. J., & Martínez-Ávila, M. (2019). Environmental performance and green culture: The mediating effect of green innovation. An application to the automotive industry. *Sustainability, 11*(18), 4874. https://doi.org/10.3390/su11184874

Gherardi, S., & Laasch, O. (2022). Responsible management-as-practice: Mobilizing a posthumanist approach. *Journal of Business Ethics, 181*(2), 269–281. https://doi.org/10.1007/s10551-021-04945-7

UN Global Compact Office and Business for Social Responsibility (2010) *Supply chain sustainability a practical guide for continuous improvement*. Retrieved June 1, 2023, from https://unglobalcompact.org/library/205

Gond, J. P., Igalens, J., Swaen, V., & El Akremi, A. (2012). The human resources contribution to responsible leadership: An exploration of the CSR–HR interface. *Responsible Leadership, 98*, 115–132.

Hadjikhani, A., Lee, J. W., & Park, S. (2016). Corporate social responsibility as a marketing strategy in foreign markets. *International Marketing Review, 33*(4), 530–554. https://doi.org/10.1108/IMR-03-2014-0104

Hotten, R. (2015). *What is Volkswagen accused of*? Business Reporter, BBC News. https://Www.Bbc.Com/News/Business-34324772

Hühn, M. P., Habisch, A., Hartman, E. M., José, G., & Sison, A. (2020). Practicing management wisely. *Business Ethics: A European Review, 29*, 1–5.

Jamali, D. (2008). A stakeholder approach to corporate social responsibility: A fresh perspective into theory and practice. *Journal of Business Ethics, 82*, 213–231. https://doi.org/10.1007/s10551-007-9572-4

Keeble, B. R. (1988). The Brundtland report: 'Our common future.' *Medicine and War, 4*(1), 17–25.

Kleyn, N., Abratt, R., Chipp, K., & Goldman, M. (2012). Building a strong corporate ethical identity: Key findings from suppliers. *California Management Review, 54*(3), 61–76. https://doi.org/10.1525/cmr.2012.54.361

Küçükoğlu, M. T. (2012). Etik değerler ve etiğin kurumsallaşması. *Hukuk ve İktisat Araştırmaları Dergisi, 4*(1), 181.

Kuenzi, M., Mayer, D. M., & Greenbaum, R. L. (2020). Creating an ethical organizational environment: The relationship between ethical leadership, ethical organizational climate, and unethical behavior. *Personnel Psychology, 73*(1), 43–71. https://doi.org/10.1111/peps.12356

Kumar, R. S., & Sıngh, A. (2013). A study on the impact of unethical corporate practices on corporate brand equıty. *International Journal of Research in Commerce and Management, 4*(5), 1–6. http://ijrcm.org.in/

Lauring, J., & Thomsen, C. (2009). Collective ideals and practices in sustainable development: managing corporate identity. *Corporate Social Responsibility and Environmental Management, 16*(1), 38–47. https://doi.org/10.1002/csr.181

Leetaru, K. (2019) The problem isn't Cambridge analytica: it's facebook. Retrieved June 17, 2023, from Forbes, https://www.forbes.com/Sites/Kalevleetaru/2018/03/19/The-Problem-%C4%B0snt-Cambridge-Analytica-%C4%B0ts-Facebook/?sh=4b02154d69bf

Litwin, G. H., & Stringer, R. A. (1968). *Motivation and organizational climate*. Harvard University.

Liu, L., & Zhao, L. (2019). The influence of ethical leadership and green organizational identity on employees' green innovation behavior: The moderating effect of strategic flexibility. In *IOP conference series: earth and environmental science*, Vol. 237, Issue 5. IOP Publishing.

Lockwood, N. (2009). *HRM research quarterly business ethics: The role of culture and values for an ethical workplace*. Retrieved June 17, 2023, from http://www.shrm.org/Research/Articles/Documents/09-0711_RQ_4_2009_FINAL.pdf

Lu, H., Xu, W., Cai, S., Yang, F., & Chen, Q. (2022). Does top management team responsible leadership help employees go green? The role of green human resource management and environmental felt-responsibility. *Corporate Social Responsibility and Environmental Management, 29*(4), 843–859.

Marcouse, I. (2014). Shareholders' interests are our own. In *The Business Book* (p. 131).

Meredith, S. (2018). Facebook-Cambridge analytica: A timeline of the data Hijacking Scandal. CNBC. Archived From The Original On October 19, 2018. Retrieved June 8, 2023, from https://www.cnbc.com/2018/04/10/facebook-cambridge-analytica-a-timeline-of-the-data-hijacking-scandal.html

Norton, T. A., Zacher, H., & Ashkanasy, N. M. (2015). Pro-environmental organizational culture and climate. In Robertson, J. L., & ve Barling, J. (Eds.). *The psychology of green organizational behaviour* (pp. 322–350). Oxford Press.

Obrad, C., & Gherhes, V. (2022). A human resources perspective on responsible corporate behaviour. Case study: The multinational companies in Western Romania. *Sustainability, 10*(3), 726. https://doi.org/10.3390/su10030726

Omidi, A., & Dal Zotto, C. (2022). Socially responsible human resource management: a systematic literature review and research agenda. *Sustainability, 14*, 2116. https://doi.org/10.3390/su14042116

Pereira, T. H. M., & Martins, H. C. (2021). People, planet, and profit: a bibliometric analysis of triple bottom line theory. *Journal of Management and Sustainability, 11*, 64.

Pham, H., Pham, T., & Dang, C. N. (2022). Assessing the importance of transformational leadership competencies and supply chain learning to green innovation: construction practitioners' perspectives. *Construction Innovation, 22*(4), 1138–1154. https://doi.org/10.1108/CI-03-2021-0037

Powell, S. M., Davies, M. A., & Norton, D. (2013). Impact of organizational climate on ethical empowerment and engagement with Corporate Social Responsibility (CSR). *Journal of Brand Management, 20*, 815–839. https://doi.org/10.1057/bm.2013.14

Rushton, K. (2002). Business ethics: a sustainable approach. *Business Ethics: A European Review, 11*(2), 137–139. https://doi.org/10.1111/1467-8608.00269

Schaubroeck, J. M., Hannah, S. T., Avolio, B. J., Kozlowski, S. W., Lord, R. G., Treviño, L. K., Dimotakis, N., Peng, A. C. (2012). Embedding ethical leadership within and across organization levels. *Academy of Management Journal, 55*(5), 1053–1078. https://doi.org/10.5465/amj.2011.0064

Schein, E. H. (2010). *Organizational culture and leadership* (3rd ed.). Jossey Bas.

Selvarajan, T. T. (2006). A cognitive processing model for assessing ethical behavior of employees. *Journal of American Academy of Business, Cambridge, 9*(1), 86–92.

Şengüllendi, M. F., Bilgetürk, M., & Afacan Fındıklı, M. (2023). Ethical leadership and green innovation: the mediating role of green organizational culture. *Journal of Environmental Planning and Management*, 1–22. https://doi.org/10.1080/09640568.2023.2180347

Shafique, I., Ahmad, B., & Kalyar, M. N. (2020). How ethical leadership influences creativity and organizational innovation: Examining the underlying mechanisms. *European Journal of Innovation Management, 23*(1), 114–133. https://doi.org/10.1108/EJIM-12-2018-0269

Shin, Y., Sung, S. Y., Choi, J. N., & Kim, M. S. (2015). Top management ethical leadership and firm performance: Mediating role of ethical and procedural justice climate. *Journal of Business Ethics, 129*, 43–57. https://doi.org/10.1007/s10551-014-2144-5

SHRM/Ethics Resource Center. (2008). Business ethics: The role of culture and values for an ethical workplace, research quarterly, Published by the Society for Human Resource Management.

Smith, A. (2023). There's an open secret about cambridge analytica in the political world: It doesn't have the 'secret sauce' it claims. *Business Insider*. Retreived June 1, 2023, from https://www.businessinsider.com/cambridge-analytica-facebook-scandal-trump-cruz-operatives-2018-3

Steele, G. (2019) Setting the right sustainability goals. *Forbes*. https://www.forbes.com/sites/forbesbusinesscouncil/2021/05/27/setting-the-right-sustainability-goals/?sh=6b2df1f65ed3
The US Department Justice Report (2016). Retrieved June 01, 2023, from https://Www.Justice.Gov/Opa/Speech/Deputy-Attorney-General-Sally-Q-Yates-Delivers-Remarks-Press-Conference-Announcing-147
UN (1987). https://sustainabledevelopment.un.org/content/documents/5987ourcommon-future.pdf
UN (2005) *Report of the economic and social council for 2005*. Retrieved June 02, 2023, from https://digitallibrary.un.org/record/558051
Van der Wal, Z., & Demircioglu, M. A. (2020). More ethical, more innovative? The effects of ethical culture and ethical leadership on realized innovation. *Australian Journal of Public Administration, 79*(3), 386–404. https://doi.org/10.1111/1467-8500.12423
Vodafone (2015). Retrieved November 08, 2015 from http://Www.Vodafone.Com.Tr/Vodafonehakkinda/Vodafone_Surdurulebilirlik_Raporu2013-2014.Pdf. https://www.vodafone.com/business/vodafone-business-ventures
Walumbwa, F. O., Mayer, D. M., Wang, P., Wang, H., Workman, K., & Christensen, A. L. (2011). Linking ethical leadership to employee performance: The roles of leader-member exchange, self-efficacy, and organizational identification. *Organizational Behavior and Human Decision Processes, 115*(2), 204–213. https://doi.org/10.1016/j.obhdp.2010.11.002
Wang, C. H. (2019). How organizational green culture influences green performance and competitive advantage: The mediating role of green innovation. *Journal of Manufacturing Technology Management, 30*(4), 666–683. https://doi.org/10.1108/JMTM-09-2018-0314
Wang, J., & Feng, T. (2022). Supply chain ethical leadership and green supply chain integration: A moderated mediation analysis. *International Journal of Logistics Research and Applications*, 1–27. https://doi.org/10.1080/13675567.2021.2022640
Wood, B. P., Eid, R., & Agag, G. (2021). A multilevel investigation of the link between ethical leadership behaviour and employees green behaviour in the hospitality industry. *International Journal of Hospitality Management, 97*, 102993. https://doi.org/10.1016/j.ijhm.2021.102993
Yang, Z., Sun, J., Zhang, Y., & Wang, Y. G. (2017). Green, and it's green: A triad model of technology, culture, and innovation for corporate sustainability. *Sustainability, 9*, 1369.
Zhang, Q., & Ma, Y. (2021). The impact of environmental management on firm economic performance: The mediating effect of green innovation and the moderating effect of environmental leadership. *Journal of Cleaner Production, 292*, 126057. https://doi.org/10.1016/j.jclepro.2021.126057

Responsible Human Resource Management: A Strategic Approach

Santiago Gutiérrez-Broncano, Mercedes Rubio-Andrés, Pedro Jiménez-Estévez, John Opute, and Camilo Giraldo-Giraldo

Abstract This chapter first addresses the definition of human resources as a key strategic resource for organisations and the importance of the human resources department as responsible for implementing responsible personnel management policies, assessing its functions and objectives. Based on the identification of socially responsible practices towards employees, it is considered whether they also have a positive impact on financial performance. Special emphasis is placed on high-performance practices because they are linked to responsible human resource management. It is also important that companies communicate adequately and externally their human resource policies based on good practice, on a voluntary or mandatory basis to comply with legal requirements. This also allows the most responsible companies with respect to their employees to form part of rankings, such as "The best place to work" or "Merco empresas responsables". The chapter covers codes of ethics as a commitment that responsible employers make to their employees, managers and how they should act in the organisation. The chapter also includes some examples of problems in the field of human resource management.

Keywords Responsible human resource management · Responsible companies · Responsible management · Strategic personnel management

S. Gutiérrez-Broncano (✉)
Faculty of Social Sciences and Information Technologies, Business Administration Department, University of Castilla-La Mancha, Talavera de la Reina, España
e-mail: santiago.gutierrez@uclm.es

M. Rubio-Andrés
Faculty of Comerce, Complutense University of Madrid, Madrid, Spain

P. Jiménez-Estévez
Faculty of Law and Social Sciences, Business Administration Department, University of Castilla-La Mancha, Toledo, Spain

J. Opute
Innovation, Leadership, Strategy and Management Department, London South Bank University, London, UK

C. Giraldo-Giraldo
Faculty of Law and Social Sciences, University of Castilla-La Mancha, Toledo, Spain

C. Machado (ed.), *Building the Future with Human Resource Management*,
Management and Industrial Engineering, https://doi.org/10.1007/978-3-031-52811-8_6

Summary

Given the growing importance of the social vision in business management, more and more companies are advocating a socially responsible orientation in the management of their human resources with the intention of improving the quality of life and well-being of their employees, also aware that this orientation generates a positive impact on the value and results of the business as well as increasing their competitive strategy.

- To define what it means to be a socially responsible company and what practices should be carried out.
- To understand what the HR departments of socially responsible companies are like.
- To identify the most relevant issues in terms of human resources practices under social responsibility.
- Reflect on the socially responsible initiatives and challenges that companies must face to improve human resources management processes.
- Importance of external communication of socially responsible HR actions.
- To assess the most relevant rankings of the companies that are most committed to their employees and the repercussions on their financial results.

1 Introduction

When Freeman (1984) published the stakeholder theory, he defined the importance of stakeholders in companies and in decision-making, moving from an approach where only business management had to attend to the interests of stakeholders (Friedman, 1970) to one where workers, suppliers, customers, competitors, and society in general have a fundamental impact on companies. Due to the great importance of stakeholders in organisations, companies are obliged to carry out responsible management, incorporating what we know today under the term corporate social responsibility (CSR). Companies that want to be successful in the medium and long term must identify the full range of stakeholder needs and be able to build lasting relationships with them (Dmytriyev et al., 2021).

Despite its great importance, most of the research we have found on CSR has focused on analysing how policies affect the perceptions of external stakeholders such as customers, governments, markets in general, suppliers, etc. (Del-Castillo-Feito et al., 2021, 2022; Toussaint et al., 2021). In contrast, the internal sphere, and the impact of these actions mainly on intangible assets remain relatively unexplored (Blanco-González et al., 2020).

In this chapter, of the three dimensions into which CSR has traditionally been divided: economic, social, and environmental, we are interested in addressing all the actions that revolve around workers and which would be included in the social dimension, being those measures relating to how the company manages its human resources, seeking work-life balance, equality, communication, promotion of fluid relations between employees and managers, degree of consideration of the interests of employees, participation in the company and occupational health (Hourneaux et al., 2018). Companies that implement these social responsibility initiatives towards their employees will generate trust within the company (Blanco-Gonzalez et al., 2020; Scherer et al., 2013) and improve their financial performance (Orlitzky, 2013; Wang et al., 2016).

We define socially responsible companies as those that respond to a management style based on sustainable development, described as development that meets the needs of today's world without compromising the ability of future generations to meet their own needs (Marcelo et al., 2012). Being socially responsible does not only mean fully complying with legal obligations, but also going beyond compliance by investing "more" in human capital, the environment and stakeholder relations as stated in the Green Paper (Commission of the European Communities, 2001). This implies the application of higher standards than those required by social legislation, for example in terms of training, working conditions or management-employee relations, which, according to the Green Paper, can even have a direct impact on productivity. This new dimension opens a way to manage change and reconcile social development with increased business competitiveness.

2 Human Resources: Definition and Processes of Personnel Management

Human resources are defined as each person working in the organisation, regardless of their level in the organisational hierarchy. They are the most valuable resource in a company because, unlike other types of resources (e.g., facilities, equipment, technology, etc.), they cannot be copied or replaced.

The right employees play a key role in the success of organisations due to their presence in the day-to-day execution of operational strategies (Barrena-Martínez et al., 2019). In short, they are considered the most important source of competitive advantage for companies and, therefore, their management is no longer perceived as a cost but as a fundamental investment for their future (Olcese et al., 2008).

Following a sustainable and socially responsible approach to human resource management will improve employee performance and engagement, and lead to better organisational results (Guo et al., 2017; Toussaint et al., 2021). The successful implementation of different CSR policies is highly correlated with business outcomes, to the extent that socially responsible human resource management is implemented (Shen & Benson, 2016).

We can summarise the policies on which HR decisions are made as follows:

Human resources policies	Questions for reflection on responsibility
Requirements planning	Are staffing needs planned methodically, will need or availability be higher? If the need is greater than the availability, how will the recruitment process be carried out? If it is less, how will the break in employment be carried out, and will alternatives to dismissal be used?
Workflows	Is the division of labour in the organisation efficient and accountable, is teamwork correct, are the responsibilities assumed by employees commensurate with the work assigned? Is delegation taking place?
Incorporation into the company	how are management positions filled, internally or externally, how are new vacancies managed, are fair and objective criteria considered in the selection process, is there an induction programme for new employees? Are fair and objective criteria considered in the selection process, and is there an induction programme for new employees?
Remuneration	Are all employees performing similar functions remunerated equally, is internal equity? And external equity - is the value contribution of employees considered?
Performance assessment	Are appraisals carried out, do employees also rate their managers, are rewards associated with them if they are positive, are tools used to ensure their neutrality?
Formación, desarrollo y carrera profesional	Is the company committed to improving the training of its employees, does it consider the preferences of employees in their career development, is it an important investment for the company, is it an important investment for the company?

3 The Human Resources Department in Socially Responsible Companies

The world of work and people management is rapidly evolving, and the way people are managed globally is extremely important for business efficacy. There is fierce competition at global level than before. As a result, more companies are operating at global level than before. More companies are making the same products with increased pace of change impacting on technology, customer requirement, labour market changes, political geography, and varying events (unexpected or otherwise).

In essence, the direction and scope of an organisation over the long term which assures business efficacy defines the strategy and enhances same. There is substantial evidence that the patterns of managing people from an international perspective have been greatly influenced by the dynamics of the strategic direction of the business. There is widened scope of management and strategic decisions culminating into strategic choices, strategic influences and strategic opportunities which is being further discussed in this chapter.

A very relevant issue is the role of the human resources department in responsible companies. We can state that the human resources department has undergone considerable evolution over the years. In the 1950s, its purpose was purely administrative, aiming at controlling employees, keeping records, and paying salaries. With the Workers' Statute (1980) and the first collective bargaining agreements, it began to have greater autonomy, becoming a staff department in the 1990s (De la Calle and Urbina-Criado, 2018). Nowadays, it is more commonly positioned as a transversal department in which all departments and people are involved (Cascio, 1998; Gómez-Mejía et al., 1999; Navarro-Abal and Climent-Rodrígez, 2012), with a completely different vision.

The HR departments of responsible companies have a strategic vision where the human resource is the intellectual capital of the company. Their objective is to attract, retain and motivate the most valuable employees for the organisation, referred to as "talent", through attractive human resources policies, such as training and performance management, professional careers, fair and people-based remuneration management, work-life balance, etc. These issues will be addressed in the following sections.

Ultimately, the hierarchical dependence and powers of the human resources department are very relevant. When it becomes part of top management decisions, policies based on good practice have more room for manoeuvre.

Objectives of the HR department from the perspective of social responsibility (Fig. 1):

- Ensure compliance with current regulations and legislation.
- Propose measures for attracting and retaining talent.
- Generate a healthy working environment
- Seek internal and external equity in salary remuneration.
- To train employees adequately and continuously and to establish attractive career paths
- Working optimally with functional managers
- Provide employees with the right conditions to perform their activities
- Establishing a fair work distribution and workload commensurate with the working hours of employees
- Promoting transparency in all human resource management processes

4 Socially Responsible Practices: Challenges and Challenges

Finding a universal definition of responsible human resource management could be a daunting task, as this concept is rooted in ethical imperatives and is therefore highly contextual (Lähteenmäki & Laiho, 2011). In this paper, we argue that socially responsible practices do not simply seek to provide employees with good working conditions based on legal requirements and regulations (e.g., minimum wage), but

The "All Together" Plan at Nueva Pescanova

GRUPO NUEVA PESCANOVA

1. Provide the organisation with the tools, processes and systems that allow it to function as a globally integrated organisation in the HR area, including: flexible remuneration, integration of global payroll, redesign of the HR area, etc.
2. Connect with the heart of your people: improving the candidate experience, local crew development, future CEO development programmes, internal mentoring, development of female managers, excellence awards, etc.
3. Build the next generation of Pescanova leaders, through an internal development programme that can translate into better professional performance.

Fig. 1 Example of human resources objectives set by the company Pescanova

rather proactively seek to improve employees' work experiences and meet their personal and social expectations in an ethical and socially responsible manner (Greige Frangieh & Khayr Yaacoub, 2019), thus contributing to employee well-being and social demands (Heikkinen et al., 2021).

Business management, through its human resources department, establishes how it will lead and manage its people, which translates into a series of concrete policies, which should be based on social responsibility criteria. This differentiates CSR from other measures aimed solely at improving corporate image.

To attract and retain talent, it is essential for companies to implement responsible human resources policies. Socially responsible actions generate a good working environment and achieve higher levels of employee satisfaction, which means that employees want to pursue their professional careers, reducing employee turnover and even attracting new candidates who will consider working in these companies as one of their options.

The recognition of the contemporary issues that impact on the strategic direction of the international HRM function are crucial for company efficacy. These measures include Corporate Social Responsibility, Cultural Intelligence, Emotional Intelligence, Covid 19, and Well-being/Work-life-balance. The extent to which these are managed in the respective regions across the globe are dependent on various organisational/contextual issues such as history, size, trade group and the labour market orientation. Furthermore, the extent to which most of these contemporary issues are embraced are dependent on whether the business is driven by the desire to be an 'employer of choice' and conscious of the public relations image that is valuable/beneficial to employer branding. The chapter thus discusses the impact of various contemporary issues and the applicable benefits. It highlights the capacity to communicate, work effectively, recognising the impact of culturally diverse situations and environmental capital to the contribution of the company's success.

Socially responsible practices towards employees can vary enormously, and there is no total unanimity on which human resources policies or practices are socially responsible. However, it is clear in the professional sphere that including CSR in human resources management can have a positive and significant influence on the degree of employee commitment, sense of belonging to the company, job satisfaction and motivation (Barrena-Martínez et al., 2012). The key is that the initiatives adopted are the organisational aspect of CSR policies and plans that should be extended to all levels and areas of management, and not be reduced to the sporadic and subordinate activities of a department or a specialist.

For Forética (2006), responsible companies require fair and respectful treatment of employees. In addition to respect for the constitutional, contractual, and collective rights of workers, responsible human resource management implies a commitment to the promotion of non-discrimination and merit, to the reconciliation of work and family life, and to a pleasant and positive working environment. Employees should have clear expectations about their future, according to objective criteria and without fear of unjustified pressure, harassment, or exclusion. Other common features of CSR in this field are the integration of workers with disabilities, the explicit elimination of gender discrimination and the commitment to prevent and prosecute all forms of abuse and harassment at work.

Therefore, as we have defined, the way of leading and managing people is what will determine whether companies are within the socially responsible sphere, through compliance with labour legislation, the implementation of better working conditions than the mandatory minimum, the search for the integration of employees in companies, the development of professional careers in accordance with the needs and expectations of employees, the conciliation of professional and personal life, etc.

In short, these practices, which constitute the challenges faced by companies, could be considered to improve the quality of work and the quality of life of employees.

- Attracting and retaining employees.
- Work-life balance management.
- Training and career development of employees.
- Diversity and equal opportunities.
- Encourage all platforms of employee voice.
- Transparency, communication, and dialogue with employees.

- Occupational health and safety prevention.
- Fair remuneration and social benefits.

The GRI standards for sustainability reporting include five categories:

1. Employee-management relations: includes policies to encourage dialogue and information flows with employee representatives, provides transparent information on remuneration systems and develops work-life balance plans.
2. Health, safety and social benefits: includes occupational risk prevention programmes, develops regular checks on health and safety conditions at work, trains employees on these issues and includes plans or benefits complementary to the standard health risk coverage.
3. Employee training: develops ongoing training programmes to promote continuous employee knowledge and performance improvement, develops training activities that enable employees to adapt to technological and organisational changes and/or reduce the risks of exclusion from the workplace, and offers environmental training activities to its employees.
4. Diversity, opportunities and non-discrimination: supports diversity in terms of gender, age and social class, accepts flexible working relationships to adapt to the situations of its employees (age, sex, disability) and offers work opportunities to young people and/or people with disabilities.
5. Corporate volunteering: supports paid volunteering programmes for its employees and encourages employee participation.

There are also some studies in the literature that focus on more concrete practices that have been less studied in the past. For example, Lombardi et al. (2020) recommend adopting more durable employment contracts to make work safer for employees, Obrad and Gherhes (2018) point out that facilities should be provided for people with disabilities and for telecommuting, and Celma et al. (2018) and Lin-Hi et al. (2019) insist on designing non-discrimination policies in the workplace.

5 High-Performance Practices as Responsible Management

High Performance Work Systems practices are those human resource actions that seek to strengthen the knowledge, skills, and abilities of employees, increase their motivation and improve the retention of qualified employees (Huselid, 1995). They are human resource management practices that create value.

They seek to foster workers' commitment to the organisation and its objectives, emphasise autonomy and self-control over their work (Guest, 1987), generic training (Pfeffer, 1994) and achieve greater employee involvement in the company (Sanz and Sabater, 2002).

Several authors have identified high performance practices. These practices can be summarised as follows (Gutiérrez-Broncano et al., 2010; Martí, 2008).

- Broad job design: Broad jobs, low vertical, and horizontal specialisation.
- Selective recruitment: Defined recruitment and selection criteria in line with job requirements. Past, present, and future behaviour is analysed.
- Fixed contracts: Stability in the job, avoiding temporary employment.
- Extensive training: Great importance and investment in employee training and development. Continuous and evaluated training for all employees. Incentives for new skills
- Performance appraisal: Results-based appraisal systems.
- Contingent compensation: Salary policies focused on individual and group results. Remuneration above market average. Tangible and intangible, flexible, participative, and public incentives.
- Employee autonomy: Granting the job and the employee him/herself great autonomy to determine their own conduct and standards at work. Decentralisation and less vertical specialisation.
- Internal promotion: Great opportunities for professional development in the company through a formal career system.
- Teamwork: Incentive and encouragement of teamwork, high commitment among members, and low turnover.
- Information sharing: Information systems shared and open to all members of the organisation.
- Job security: High level of job security. Specialised personnel to improve safety at work.

A large part of these practices has been implemented in the company Semco, which we use as a case study. Semco is an example of a pioneering company in responsible human resource management. Semco is a company that has awakened an unusual interest in the world for its innovative practices in the responsible management of its resources. This has led it to become the largest marine and food processing machinery manufacturing company in Brazil. A Brazilian employability survey found that more than 2,000 people said they would accept any job if they wanted to work for Semco. Another survey of new university graduates found that 25% would want to work there.

The company's founder is Ricardo Semler, who has earned Semco recognition for its innovative and responsible management initiatives. We highlight the following actions by the company:

- Weekly working hours. At Semco, there are no strict rules regarding working hours or days spent in the office. All employees work as if they were "on leave" every day of the week (Semler, 2004).
- Working hours. Most employees set their own working hours (Rubio-Andrés et al., 2015). They also decide what their start time is (usually from 7 to 9 am), and although the working day is set at 8 h, there is no control over it, with several occasional breaks.

- Remuneration. More than 150 people on the board set their own salaries and bonuses. Each division has an individual profit-sharing programme. 23% of the profits are distributed among the employees according to the criteria determined by each group. The medium-term associates earn higher salaries than the coordinators and partners.
- Participation in the results. If the employees are involved in the success of the company, it is logical for the management to share in the company's results. Regardless of the position held.
- Own decision-making. Employees can make their own decisions without the approval of management. They have the freedom to set their own production quotas.
- Self-monitoring. To monitor the achievement of the targets, the workers took the initiative to set up scoreboards in the factories in order to see whether they were achieving the set production.
- Self-motivation. For Semco, it is not only the inclusion in the area that is important, but also the integration in the company. The motivation to work must come from the employees, who are the ones who take the initiative in many decisions, so we can say that motivation comes "from within".
- Continuous training. To avoid routine in the workplace, in addition to possible job rotation, Semco considers the development of the professional career in an upward direction, without the level of studies achieved so far being a handicap. For workers from the operational base to become financial or technical managers, if they have sufficient skills, the company trains them and pays for their studies.
- Evaluation of workers. Twice a year, each employee completes a questionnaire to evaluate their managers.
- Recruitment based on responsibility. The recruitment process is based on each employee's capacity for responsibility. They are not looking for people who work only for the salary, and who do not want to work hard for the company. This is in response to the treatment of the employee because the employee acts in a responsible manner.
- Employee job satisfaction. Semco believes that "success is not only measured in profits and business growth, but in the balance achieved in the lives of its employees" (Semler, 2004), because when people are balanced, they are more prepared to work towards and achieve the company's goals, merging their own personal aspirations with those goals.
- Job security. At Semco, job security is guaranteed through permanent contracts and turnover is very low.

More information can be found by watching the following video: https://www.ted.com/talks/ricardo_semler_how_to_run_a_company_with_almost_no_rules?language=es.

6 Responsible Human Resources Management and Its Business Impact

One question that companies ask and which applied research is trying to answer is whether those that are more socially responsible achieve higher financial performance. In general, we can affirm that the most prominent studies affirm that there is an impact between those companies that are more responsible and financial performance (Margolis and Walsh, 2001; Orlitzky, 2013).

Olcese et al. (2008) cite the study conducted by SAM and the Boston Consulting Group, which concluded that there is a positive correlation between human capital management and financial performance. According to these authors, companies that manage their human resources properly reduce their costs and better exploit their innovative potential, thereby generating higher revenues and achieving higher rates of return on investment. These companies are successful in attracting and retaining the right employees and anticipate their changing expectations regarding issues such as job content, flexible working arrangements, fringe benefits, training, and career development.

In Paauwe and Richardson's (1997) model, a relationship is established between human resource management policies, which are based on responsible management, and the positive results achieved in human resources and financial performance (Fig. 2).

In addition to finding evidence that responsible HR practices increase business performance (Buciuniene & Kazlauskaite, 2012; Chanda & Goyal, 2020; Luu, 2020), these practices also improve companies' CSR performance (Buciuniene & Kazlauskaite, 2012), intellectual capital (Barrena-Martinez et al., 2019) and company reputation (Sobhani et al., 2021; Ramos-Gonzalez et al., 2022), and even innovation (Ramos-Gonzalez et al., 2022, 2019) and firm reputation (Sobhani et al., 2021; Ramos-Gonzalez et al., 2022), and even innovation (Ramos-Gonzalez et al., 2022).

Fig. 2 Relationship between responsible HR policies and business performance. Fuente: Paauwe and Richardson (1997)

Furthermore, Low and Bu (2021) suggest that responsible HR practices can help the organisation to digitise work processes in a more engaged way, while Parkes and Davis (2013) argue that organisations could attract talent faster by applying these responsible practices.

Other studies confirmed that responsible human resource management contributes to improved employee engagement (Chanda & Goyal, 2020; Diaz-Carrion et al., 2019; Jamali et al., 2020; Low & Bu, 2021), civic engagement (Jamali et al., 2020; Lee, 2021; Sobhani et al., 2021; Zhao & Zhou, 2021), task performance (Chang et al., 2021), employee satisfaction (Chanda & Goyal, 2020; Sorribes et al., 2021), well-being (Diaz-Carrión et al., 2019), loyalty (Brown, 2018), employees' propensity to identify with their organisation (Jamali et al., 2020; Sánchez-Hernández et al., 2020), improved employee-employer relationships (Lee, 2021), employee trust in the company (He et al., 2021; Sorribes et al., 2021), empathy (Shao et al., 2019), knowledge sharing (Jia et al., 2019), innovation (Revuelto-Taboada et al., 2021) and intrapreneurial behaviour (Giang & Dung, 2021; Luu, 2020). In addition, it has also been shown that these responsible practices can help employees make sense of their jobs in a meaningful way (Luu, 2021) and even make employees more resilient in times of crisis, such as the recent COVID-19 pandemic (He et al., 2021). Finally, we found studies that conclude that companies that implement such practices can decrease employees' intention to leave the company (Sobhani et al., 2021).

7 External Communication of Responsible HR Management

Good practice with respect to human resources management is not only an internal matter but can also be disseminated so that it is known by the other stakeholders involved in or affected by the business activity: customers, investors, and society in general, which will improve its reputation or its external image. In this sense, it is very important how companies communicate their responsible practices related to their human resources, whether in an informal way, such as through their web pages or publications of interviews with their managers, etc., or above all through formal communication, through the preparation of non-financial or sustainability reports, the GRI reports, which are subject to external audits and therefore verifiable in terms of their content. In both cases, it can be seen as an opportunity to communicate the responsible management of human resources.

In the following, we set out in more detail what the various reports cited above contain and what information related to responsible human resources policies they contain.

7.1 Non-Financial Reporting: Law 11/2018

The mandatory presentation by companies of the non-financial report is quite new in some countries, such as Spain, for example. The quest to improve transparency and sustainability for investors, consumers and society in general has been regulated through Law 11/2018 on non-financial information and diversity, whose precedent is Royal Decree Law 18/2017, both to comply with the European Directive (2014/95/EU).

Although the contents to be reflected in the non-financial report vary in nature, in relation to human resources, the regulations establish that companies that meet the requirements must report on the following:

- number of employees according to diversity criteria (gender, age, country, or professional category)
- remuneration and pay gap
- average remuneration of directors and managers
- absenteeism data and number of dismissals
- organisation of working time (absenteeism, work/life balance, work disengagement…)
- health and safety in the workplace (accidents at work and their frequency and seriousness, occupational illnesses broken down by gender, etc.),
- employee training (policies implemented and data on the total amount of training hours provided).
- equal treatment and opportunities, protocols against harassment and discrimination.
- prevention of risks of violation, elimination of forced labour and child labour.

7.2 The GRI Indicators: The Social Performance Indicators

GRI (Global Reporting Initiative) is an independent international organisation that pioneered sustainability reporting. The GRI Standards present global best practices for publicly reporting an organisation's economic, environmental, and social impacts. Sustainability reporting based on these standards provides information about the positive and negative contributions of organisations to sustainable development.

Social and human resources issues are covered in GRI 4 (social). It considers labour-management relations; freedom of association and collective bargaining, occupational health and safety, training and education, diversity and equal opportunities, non-discrimination, safety practices, etc.

It is made up of numerous indicators that are used to assess the level of compliance.

An example of the information to be provided by the company for the sustainability report related to good practice in the report according to GRI 401:

- Total number and rate of new employee hires during the reporting period, by age group, gender, and region.

- Total number and rate of employee turnover during the reporting period, by age group, gender, and region.
- Total number of employees who were entitled to maternity/paternity leave by gender.
- Total number of employees requesting parental leave, by gender.
- Total number of employees who returned to work in the reporting period after the end of their parental leave, by gender.

Ratio to be calculated:

$$\text{Retention rate} = \frac{\text{Total number of employees continuing 12 months after returning to work following a period of parental leave}}{\text{Total number of employees returning from parental leave in the previous reporting periods}}$$

7.3 *Responsible Employer Rankings*

The impact of companies' responsible actions towards their employees has been measured by various organisations and institutions, which have drawn up rankings that allow comparison between companies. The following stand out:

GREAT PLACE TO WORK

"Great Place to Work" analyses those companies that voluntarily submit themselves to an evaluation of the good practices of their organisational culture. Aspects such as salary, internal communication, respect for people, fairness, working environment, etc. are studied. With this data, a public ranking of the "best companies to work for" is drawn up (Forética, 2006). One of its advantages is that, on the one hand, professionals will be attracted to the companies that appear in the best positions in the ranking, choosing them over others that are not. The benefit is mutual, as companies will attract more candidates and will be able to choose among the best.

The "Best Workplaces" ranking rewards the best places to work. These were the companies that, thanks to their High Levels of Trust and good organisational culture practices, were part of the Ranking 2023 Spain.

More than 1,000 employees	501–1,000 employees	251–500 employees
Lilly S.A. Pharmaceutical industry **DHL Express Spain** Logistics and inter-country transport **Reale Seguros Generales** Insurance **Pelayo** Insurance **Banca March** Financial	**Salesforce** Information Technology **Admiral Seguros** Insurance **DKV** Insurance Group **Pernod Ricard España** Food and beverage **Abbvie** Pharma	**Cisco** Telecommunications **Stryker** Medical Service **Ingram** Micro Information Technology **Hilton** Hospitality **Thoughtworks** Technology

Source Great Place to Work, 2023. Informe_Best_Workplaces_2023.pdf (greatplacetowork.es)

MERCO

Merco (Business Monitor of Corporate Reputation) is one of the world's benchmark monitors. Merco currently produces eight monitors, two of which consider responsible human resources policies that are positively valued, and which imply that they have a high reputation: Merco Responsibility and Corporate Governance (the most responsible companies) and Merco Talent (the companies that best attract and retain talent).

In the Merco Responsibility and Corporate Governance ranking, in terms of responsibility towards employees, talent management programmes, work-life balance programmes and management of diversity and equal opportunities are valued.

Merco Talent considers the organisation and resources available for work, the people management carried out in the company, the training offered by the company, the working environment, the ability to retain good professionals and the functioning of internal communication.

8 Codes of Conduct for Employees

A code of conduct is a management tool through which a company communicates to all its members the behaviour that they must follow in the company and the relations with the company's different stakeholders (Olcese et al., 2008). They have been increasingly incorporated into the management of companies to respond to the needs posed by the environment (Sánchez Ruíz-Lozano et al., 2016). The shared values of a company act as the conscience of the organisation and allow the decentralisation of decision-making. In this sense, a code of conduct makes it possible to specify what the organisation expects from its members in different issues or situations in which they must act and make decisions.

It is necessary for a company's managers and workers to have a guide that tells them more explicitly what they must do, or not do, in certain circumstances. It establishes the guidelines to be followed and is therefore of great importance in the process of internalisation of the values of a company by all its components (Olcese et al., 2008) and establishes aspects relevant to the responsible management of human resources, such as equal opportunities, health and safety at work, diversity, the search for a good working environment, work-life balance, professional development, etc.

Web Search

Below, we propose some links where you will find the codes of ethics and conduct available on company websites. Consult them and evaluate what proposals are made in terms of human resources and whether they seem to you to be socially responsible companies, assessing whether they are sufficient.

Banco Santander: https://www.santander.com/csgs/Satellite/CFWCSancomQP01/es_ES/pdf/Codigo_General_de_Conducta_300518.pdf

Repsol: https://imagenes.repsol.com/es_es/RepsolCodeOfEthicsAndBusiness_tcm7-741874.pdf

Pepsico: https://pepsico.es/pdf/2016-es-euro-spanish-booklet-code-of-conduct-a4-es.pdf

NH Hoteles: https://www.nh-hoteles.es/corporate/sites/default/files/files-rsc/ia-comp-spa-300616-cdec.pdf

Burger King España: https://www.bkspain.es/resources/pdfNews/codigoConductaBKSpain.pdf

9 New Challenges in the Field of HR Management

Inadequate human resource management has a negative impact on business success, affecting both the individual and the organisation, and having an impact on society and future customer demand. One of the main challenges facing personnel management resulting from ineffective human resource policy design is demotivation, which at an extreme level can result in the employee suffering from "burn-out syndrome", a disease that has been typified by the WHO. The World Health Organisation (WHO) has acknowledged that "burnout syndrome" is included in the International Classification of Diseases, so companies must implement prevention plans to avoid it.

It is known that 10% of workers are currently suffering from it, especially in the financial, commercial, health and customer service sectors. And every day you are faced with a boss who does not listen, with his inability to add value to the company and his team, with the poor distribution of economic incentives or the time we spend training young talent only to let them go because they do not have a project. These are situations that produce an unbearable weariness and apathy according to most employees and managers.

Being burnt out is nothing new for hundreds of workers, 10% of the total according to experts, but it is new for those who suffer from it to recognise it. Although there are no figures on the number of sufferers or sick leave, as until now it was not considered an illness, the fact is that burnout syndrome is a ticking time bomb. It differs from stress, which causes 30% of sick leave, because it "can be beneficial and it gets you going," explains Francisco Peñalver, a specialist in occupational medicine and head of the department of safety, health, and well-being at HM Hospitales.

Feeling sad and going to work without enthusiasm can be a start. But to recognise if you are on the verge of burnout, the specialist points out some symptoms: "Getting up tired despite having slept or experiencing negative feelings towards colleagues, clients or tasks, which result in low performance and poor concentration to meet objectives. These are the clear signs of this illness, the origin of which is linked to work performance.

The syndrome finds its best breeding ground in middle management and especially in the financial and health sectors (37% of doctors are affected, according to a study by Medscape), commercial networks, customer service and anyone with emotional exposure to other people. Who is the cause? Companies and workers throw the hot potato at each other. But Ignacio Cristóbal, a professor at the University of Navarra, is clear: "It is a shared responsibility".

Source: El País-Negocios (2019).

What are the negative consequences of burn-out syndrome for companies, and what responsible measures could companies put in place to avoid it?

Irresponsible human resources practices can also affect many workers, such as unjustified and subjective dismissals, unethical recruitment and selection, etc. Some examples of this malpractice can be seen in news headlines such as the ones below:

- Iberia ordered to pay 25,000 euros for requiring pregnancy tests to hire women. Iberia has rectified its decision and will not require pregnancy tests in interviews. The airline has contracted Randstad for staff recruitment. It was detected a year ago by the Labour Inspectorate and the Balearic Ministry of Employment has fined Iberia 25,000 euros for committing a very serious offence of discrimination on grounds of sex. The infraction is because the company requested pregnancy tests to hire women, according to the 'Diario de Mallorca'. Airline sources have announced that they will stop requiring the test and that it will be up to the applicants to decide whether to inform the company if they are pregnant. According to the Consejería, the facts were detected a year ago when the Labour Inspectorate was carrying out one of the controls to fight discrimination in the workplace. The inspectors discovered that the company that Iberia had contracted for the selection of personnel, Randstad, was the one requesting pregnancy tests to find out their status. This was a nationwide practice. Source: EL MUNDO10-07-2017
- Amazon dismisses a recruitment artificial intelligence for discriminating against women. The system had been trained on the profiles of job applicants over the past 10 years. Amazon experts began building an artificial intelligence system in 2014 to review the CVs of job applicants. The aim was to mechanise the search for the best talent. This tool, based on the company's 10 years of archives, learned

that men were preferable and began to discriminate against women, as Ama-zon sources told Reuters. After learning of the error, the company decided to dispense with the system. When the company realised the error, it edited the programmes so that they no longer discriminated against women. But, according to sources consulted by Reuters, there was no guarantee that the system would not fail again or be unfair in ranking candidates. As a result, executives lost hope in the project and the Seattle-based company eventually abandoned the project. Source: EL PAIS. 12-10-2018
- The BBC reports that Syrian refugee children are working in Turkey for large textile companies. In particular, the public broadcaster has uncovered seven Syrians, the youngest aged 15, working 12-h shifts at Marks & Spencer's (M&S) main supply factory, who had been recruited by a middleman who paid them in cash just one pound (1.12 euros) an hour, below the Turkish minimum wage. Investigators have also noted that refugees from the Syrian conflict are working 12-h shifts in supplier factories dyeing jeans for Zara and Mango. According to the report, these workers were handling chemicals without even protective masks. Source: ABC, 24-10-2016

Callejo Martínez (2015), General Director of Corporate Resources at Acciona, establishes the following as the main challenges for human resources management from a strategic point of view:

1. Company culture. A strong brand and culture are essential to attract and retain the best people in each geography.
2. Talent as a competitive advantage. Equal opportunities (based on merit) and recognition are key to developing and retaining the best talent.
3. Change management in the organisation. In an environment that changes so much, the speed of adaptation to change is key to competing.
4. Making the most of technology in people management.
5. The development of extraordinarily capable human resources managers.

After this reflection, the following questions could be asked: do you consider these to be realistic challenges in human resources management in responsible companies, how should companies respond to them, will companies meet them, and will they differentiate those companies that achieve them?

10 Conclusions

In the course of this chapter, the strategic and social role of personnel management has been explored in depth. Companies play a decisive role in today's society and are agents of change, so companies that want to be competitive have to take the current context into account. Companies operate in a context in which customers, suppliers, shareholders, investors and also managers and employees demand responsible management from companies and value the fact that companies develop corporate social responsibility policies.

To be able to implement these social responsibility policies, companies need to implement responsible human resource management models. In such a way that it is through the company's employees that responsible behaviour is achieved, helping companies to achieve not only economic, but also environmental and social results.

References

Barrena-Martinez, J., López-Fernández, M., & Romero-Fernández, P. M. (2019). The link between socially responsible human resource management and intellectual capital. *Corporate Social Responsibility and Environmental Management, 26*, 71–81.

Barrena-Martínez, J., López-Fernández, M., & Romero-Fernández, P. M. (2012). El valor interno de la RSC: Implicaciones en la gestión de RRHH y el clima laboral de la empresa.

Blanco-González, A., Díez-Martín, F., Cachón-Rodríguez, G., & Prado-Román, C. (2020). Contribution of social responsibility to the work involvement of employees. *Corporate Social Responsibility and Environmental Management, 27*(6), 2588–2598.

Brown, W. S. (2018). Socially responsible entrepreneurship as innovative human resource practice. *The Journal of Applied Behavioral Science, 54*, 171–186.

Buciuniene, I., & Kazlauskaite, R. (2012). The linkage between HRM, CSR and performance outcomes. *Baltic Journal of Management, 7*, 5–24.

Callejo-Martínez, A. (2015). Los desafíos para la Dirección de Recursos Humanos en el próximo quinquenio. *Capital Humano: Revista Para La Integración y Desarrollo De Los Recursos Humanos, 28*(297), 64–74.

Cascio, W. F. (1998). *Managing human resources: Productivity, quality of work life, profits.* McGraw-Hill.

Celma, D., Martinez-Garcia, E., & Raya, J. M. (2018). Socially responsible HR practices and their effects on employees' wellbeing: Empirical evidence from Catalonia, Spain. *European Research on Management and Business Economics, 24*, 82–89.

Chanda, U., & Goyal, P. (2020). A Bayesian network model on the interlinkage between socially responsible HRM, employee satisfaction, employee commitment and organizational performance. *Journal of Management Analytics, 7*, 105–138.

Chang, Y.-P., Hu, H.-H., & Lin, C.-M. (2021). Consistency or hypocrisy? The impact of internal corporate social responsibility on employee behavior: A moderated mediation model. *Sustainability, 13*, 9494.

Cinco Días: La transparencia ya está aquí: el informe no financiero https://cincodias.elpais.com/cincodias/2018/02/06/legal/1517929970_079745.html

Comisión de las Comunidades Europeas (2001). Libro Verde. Fomentar un Marco Europeo para la Responsabilidad Social de las Empresas, 366—final.

De la Calle, M. D. C., & de Urbina Criado, M. O. (2018). *Fundamentos de recursos humanos.* Pearson Prentice Hall.

Del-Castillo-Feito, C., Díez-Martín, F., & Blanco-González, A. (2021). The effect of implementing environmental policies and employees' environmental training in multinational companies' legitimacy level in emerging countries. *Journal of Cleaner Production, 312*(20), 127817.

Del-Castillo-Feito, C., Blanco-González, A., & Hernández-Perlines, F. (2022). The impacts of socially responsible human resources management on organizational legitimacy. *Technological Forecasting and Social Change, 174*, 121274.

Diaz-Carrion, R., López-Fernández, M., & Romero-Fernandez, P. M. (2019). Evidence of different models of socially responsible HRM in Europe. *Business Ethics: A European Review, 28*, 1–18.

Dmytriyev, S. D., Freeman, R. E., & Hörisch, J. (2021). The relationship between stakeholder theory and corporate social responsibility: Differences, similarities, and implications for social issues in management. *Journal of Management Studies, 58*(6), 1441–1470.

El País-Negocios. (2019). https://elpais.com/economia/2019/06/11/actualidad/1560269922_158486.html

Forética. (2006): Cuaderno Forética 1. Responsabilidad Social Empresarial. https://foretica.org/publicacion/cuaderno-foretica-1-responsabilidad-social-empresarial/

Freeman, A. M. (1984). The sign and size of option value. *Land Economics, 60*(1), 1–13.

Friedman, M. (1970). A theoretical framework for monetary analysis. *Journal of Political Economy, 78*(2), 193–238.

Giang, H. T. T., & Dung, L. T. (2021). The effect of internal corporate social responsibility practices on firm performance: The mediating role of employee intrapreneurial behaviour. *Review of Managerial Science.*

Gómez-Mejía, L. R., Balkin, D. B., & Cardy, R. L. (1999). *Gestión de los Recursos Humanos.* Prentice-Hall.

Greige Frangieh, C., & Khayr Yaacoub, H. (2019). Socially responsible human resource practices: Disclosures of the world's best multinational workplaces. *Social Responsibility Journal, 15*, 277–295.

Guest, D. E. (1987). Human resource management and industrial relations. *Journal of Management Studies, 24*(5), 503–521.

Guo, R., Tao, L., Li, C. B., & Wang, T. (2017). A path analysis of greenwashing in a trust crisis among Chinese energy companies: The role of brand legitimacy and brand loyalty. *Journal of Business Ethics, 140*, 523–536.

Gutiérrez-Broncano, S., Rubio-Andrés, M., & Monsalve, J. N. (2010). Prácticas de recursos humanos de alto compromiso y ventaja competitiva. Un estudio sobre su implantación y la generación de valor para la empresa. Estudios financieros. *Revista De Trabajo y Seguridad Social: Comentarios, Casos Prácticos: Recursos Humanos, 333*, 141–170.

He, J., Mao, Y., Morrison, A. M., & Coca-Stefaniak, J. A. (2021). On being warm and friendly: The effect of socially responsible human resource management on employee fears of the threats of COVID-19. *International Journal of Contemporary Hospitality Management, 33*, 346–366.

Heikkinen, S., Lämsä, A.-M., & Niemistö, C. (2021). Work-family practices and complexity of their usage: A discourse analysis towards socially responsible human resource management. *Journal of Business Ethics, 171*, 815–831.

Hernández, M. I., Stankeviciute, Ž, Robina-Ramirez, R., & Díaz-Caro, C. (2020). Responsible job design based on the internal social responsibility of local governments. *International Journal of Environmental Research and Public Health, 17*, 3994.

Hourneaux, F., Jr., Gabriel, M. L. D. S., & Gallardo-Vázquez, D. A. (2018). Triple bottom line and sustainable performance measurement in industrial companies. *Revista De Gestão, 25*(4), 413–429.

Huselid, M. A. (1995). The impact of human resource management practices on turnover, productivity, and corporate financial performance. *Academy of Management Journal, 38*, 673–703.

Jamali, D., Samara, G., Zollo, L., & Ciappei, C. (2020). Is internal CSR really less impactful in individualist and masculine Cultures? A multilevel approach. *Management Decision, 58*, 362–375.

Jia, X., Liao, S., Van der Heijden, B. I. J. M., & Guo, Z. (2019). The effect of socially responsible human resource management (SRHRM) on frontline employees' knowledge sharing. *International Journal of Contemporary Hospitality Management, 31*, 3646–3663.

Lähteenmäki, S., & Laiho, M. (2011). Global HRM and the dilemma of competing stakeholder interests. *Social Responsibility Journal, 7*, 166–180.

Lee, Y. (2021). Bridging employee advocacy in anonymous social media and internal corporate social responsibility (CSR). *Management Decision, 59*, 2473–2495.
Ley 11/2018 de Información No Financiera y Diversidad (29 de diciembre de 2018).
Lin-Hi, N., Rothenhöfer, L., & Blumberg, I. (2019). The relevance of socially responsible blue-collar human resource management. *Employee Relations: The International Journal, 41*, 1256–1272.
Lombardi, R., Manfredi, S., Cuozzo, B., & Palmaccio, M. (2020). The profitable relationship among corporate social responsibility and human resource management: A new sustainable key factor. *Corporate Social Responsibility and Environmental Management, 27*, 2657–2667.
Low, M. P., & Bu, M. (2021). Examining the impetus for internal CSR practices with digitalization strategy in the service industry during COVID-19 pandemic. *Business Ethics, The Environment & Responsibility, 31*, 209–223.
Luu, D. T. (2020). The effect of internal corporate social responsibility practices on pharmaceutical firm's performance through employee intrapreneurial behaviour. *Journal of Organizational Change Management, 33*, 1375–1400.
Luu, T. T. (2021). Socially responsible human resource practices and hospitality employee outcomes. *International Journal of Contemporary Hospitality Management, 33*, 757–789.
Marcelo, J. L. M., Quirós, M. D. M. M., & Quirós, J. L. M. (2012). Performance bursátil de las empresas socialmente responsables. *Cuadernos De Economía y Dirección De La Empresa, 15*(4), 221–230.
Margolis, J. D., & Walsh, J. P. (2001). *People and profits?: The search for a link between a company's social and financial performance*. Psychology Press.
Martí Audí, N. (2008). *High commitment human resource practices in Spanish call centres. Nature, sources, and effects on voluntary turnover*. Tesis Doctoral Universitat Rovira i Virgili.
Navarro-Abal, Y. N., & Climent-Rodríguez, J. A. (2012). La gestión de los recursos humanos: dos percepciones diferentes de una misma realidad. *Trabajo: Revista andaluza de relaciones laborales,* (27), 107–123.
Obrad, C., & Gherhes, V. (2018). A human resources perspective on responsible corporate behavior. Case study: The multinational companies in Western Romania. *Sustainability, 10*, 726.
Olcese, A., Rodríguez, M. Á., & Alfaro, J. (2008). *Manual de la empresa responsable y sostenible: Conceptos, ejemplos y herramientas de la responsabilidad social corporativa o de la empresa*. McGraw-Hill.
Orlitzky, M. (2013). Corporate social responsibility, noise, and stock market volatility. *Academy of Management Perspectives, 27*(3), 238–254.
Paauwe, J., & Richardson, R. (1997). Introduction special issue on HRM and performance. *International Journal of Human Resource Management, 8*(3), 257–262.
Parkes, C., & Davis, A. J. (2013). Ethics and social responsibility—Do HR professionals have the 'courage to challenge' or are they set to be permanent 'bystanders'? *International Journal of Human Resource Management, 24*, 2411–2434.
Pfeffer, J. (1994). Competitive advantage through people. *California Management Review, 36*(2), 9–28.
Ramos-González, M. d. M., Rubio-Andrés, M., Sastre-Castillo, M. Á. (2022). Effects of socially responsible human resource management (SR-HRM) on innovation and reputation in entrepreneurial SMEs. *International Entrepreneurship and Management Journal, 18*(3), 1205–1233.
Revuelto-Taboada, L., Canet-Giner, M. T., & Balbastre-Benavent, F. (2021). High-commitment work practices and the social responsibility issue: Interaction and benefits. *Sustainability, 13*, 459.
Rubio-Andrés, M., Montoya-Monsalve, J. N., & Gutiérrez-Broncano, S. (2015). Could innovative teams provide the necessary flexibility to compete in the current context? *Cuadernos De Gestión, 15*(1), 145–163.
Sánchez-Ruiz Lozano, M., De Los Ríos Berjillos, A., & Millán Lara, S. (2016). ¿Son los códigos éticos un instrumento de gestión de la responsabilidad social? Un estudio exploratorio en Andalucía. *Intangible Capital, 2*(1), 462–486

Sanz Valle, R., & Sabater Sánchez, R. (2002): Enfoque universalista de recursos humanos. Estudio empírico de sus efectos sobre los resultados de la empresa. *Revista Europea de Dirección y Economía de la Empresa, 11*(1), 163–182.

Scherer, A. G., Palazzo, G., & Seidl, D. (2013). Managing legitimacy in complex and heterogeneous environments: Sustainable development in a globalized world. *Journal of Management Studies, 50*(2), 259–284.

Semler, R. (2004). *The seven-day weekend: Changing the way work works.* Penguin.

Shao, D., Zhou, E., & Gao, P. (2019). Influence of perceived socially responsible human resource management on task performance and social performance. *Sustainability, 11*, 3195.

Shen, J., & Benson, J. (2016). When CSR is a social norm: How socially responsible human resource management affects employee work behavior. *Journal of Management, 42*(6), 1723–1746.

Sobhani, F. A., Haque, A., & Rahman, S. (2021). Socially responsible HRM, employee attitude, and bank reputation: The rise of CSR in Bangladesh. *Sustainability, 13*, 2753.

Sorribes, J., Celma, D., & Martínez-Garcia, E. (2021). Sustainable human resources management in crisis contexts: Interaction of socially responsible labour practices for the wellbeing of employees. *Corporate Social Responsibility and Environmental Management, 28*, 936–952.

Toussaint, M., Cabanelas, P., & Blanco-González, A. (2021). Social sustainability in the food value chain: An integrative approach beyond corporate social responsibility. *Social Responsibility and Environmental Management, 28*(2), 103–115.

Wang, Q., Dou, J., & Jia, S. (2016). A meta-analytic review of corporate social responsibility and corporate financial performance: The moderating effect of contextual factors. *Business & Society, 55*(8), 1088–1121.

Zhao, H., & Zhou, Q. (2021). Socially responsible human resource management and hotel employee organizational citizenship behavior for the environment: A social cognitive perspective. *International Journal of Hospitality Management, 95*, 102749.

Ideological PC in HRM: Learning with Military

Laurentino Guimarães, Ana Paula Ferreira, and Gina Gaio Santos

Abstract This chapter aims to reflect on the possible implications of adopting ideological psychological contracts in managing people in today's organizations, in an especially difficult worldwide context. Our study aims to identify the type of Psychological Contract (PC) held by a group of Portuguese militaries under a restructuring scenario. We uncovered the existence of relational, transitional, and ideological PCs and discussed the possible implications that can be made for military organizations regarding managing people using PCs, especially ideological ones.

Keywords Psychological contract · Military context · People management · Employer-employee relationship · Restructuring

1 Introduction

The past few decades have brought significant changes in the workplace. One of the most notable shifts in labor relations that have profoundly impacted how people are managed in the workplace relates to losses in job security, as many workers now face greater uncertainty about the future of their employment.

The present study targets the psychological contract (PC) in a military context and is conducted with officers belonging to the Permanent Personnel of the Portuguese Army (PPPA) on active duty, with a permanent link with the Army and assigned to the effective service, or in conditions to be called to its performance (EMFAR).

The Military is a relatively understudied context, where research is particularly incipient and needs more context-specific data on PC types (Kraak et al., 2020; Van de Ven, 2004). Moreover, the Military organization, at the time of the research,

L. Guimarães
Higher School of Technology of Fafe and CIDI-IEES, European Institute of Higher Studies, Medelo, Portugal

A. P. Ferreira (✉) · G. G. Santos
School of Economics and Management and CICS NOVA UMinho Research Centre, University of Minho, Braga, Portugal
e-mail: aferreira@eeg.uminho.pt

C. Machado (ed.), *Building the Future with Human Resource Management*, Management and Industrial Engineering, https://doi.org/10.1007/978-3-031-52811-8_7

underwent a restructuring that justified the aims of this study. Thus, we conducted 42 in-depth interviews with a group of military officers with the following aims: (1) identify PC types that may exist in the Portuguese Army, and (2) analyze the implications of the PC typology for human resources management.

The Armed Forces, an eminently bureaucratic and highly centralized organization, is a system in which professional characteristics, especially of the officer category, change over time. According to Janowitz (1971), the military organization is made up of organizational rules and norms that include but are not limited to, the direct administration of violence. Examples of violence management situations are the humanitarian and peacekeeping missions in international settings under the determination of the United Nations (UN), especially since 1990.

The PAF is made up of three branches: Navy, Army, and Air Force. Each branch is structured into three major hierarchical categories of personnel as follows, in descending order: officers, sergeants, and enlisted personnel. Following the process of professionalization of the PAF since the late 1980s, the military members of these categories are now professionally linked to the PAF in two different ways, depending on their employment relationship: the permanent personnel, who "render professional service with a permanent contract" (Nº. 1 of Art. 113 of the EMFAR, 2015) and military personnel under contract and voluntary service, which "are holders of a fixed-term public employment contract" (Art. 267 of the EMFAR, 2015).

Of the three branches mentioned, the one with the largest staff is the Army, where its members constitute the focus under research.

2 Psychological Contract Conceptualization

The PC allows us to understand the employee-employer relationship (Dulac, 2005; Guest & Conway, 2002), especially in terms of employees' attitudes and behaviors, like commitment, and organizational citizenship behaviors (De Vos et al., 2003; Robinson & Morrison, 1995), or turnover (De Vos et al., 2003; Guest, 2004). It is distinct from the legal-formal employment contract by its unwritten nature, therefore implicit (Suazo et al., 2009), and by the idea of a relational exchange between the two parties. This exchange is based not only on the written or explicitly agreed (legal contract) but mostly on both parties' perceptions implicitly generated about the promises made to each other (Conway & Briner, 2005).

The PC can be understood as the unwritten expectations and beliefs that employees have about the terms of their relationship with the organization that are shaped over time by both parties (organizations and employees) in interaction, and it is based on expressed or implied promises (Rousseau, 1995). This definition has been relatively consensual over time (e.g. Conway & Brinner 2005; Freese et al., 2011; Vantilborgh et al., 2012). According to the literature (e.g. Bal et al., 2010; Freese et al., 2011), the PC's contents include the individuals' understanding of their employer's obligations (e.g. providing career advancement and job training, adequate pay, job security), but also of their obligations (e.g. working overtime, organizational loyalty, willingness to perform extra duties, and good job performance job in terms of quality and quantity (Bal et al., 2010; Freese et al., 2011).

Summarizing, the content of the PC translates what is exchanged, the resources made available by each party (employer-employee) within a perceived agreement, and how they can be exchanged, in the sense that one understands the precise way in which the exchange of mutually available resources will take place (Conway & Briner, 2005).

Key to organizations, the PC presents, according to Shore and Tetrick (1994) three functions: (1) reduces the insecurity about the ambiguity of the relationship between the contracting parties and not filled by the formal contract; (2) adjusts individual behavior according to the result of the comparison between the reciprocal obligations, and (3) develops feelings of individual influence over the organization, since it induces the option between fulfilling or not fulfilling their obligations.

Coyle-Shapiro and Parzefall (2008) and Rousseau (1989) consider the PC a construct that fits Blau's (1964) theory of social exchange, which considers two types of exchange relations within social processes: the economic and the social. The first is of a specified nature, usually based on a formal contract as a guarantee that the parties fulfill the specified obligations. The second includes unspecified obligations and relies mainly on trust between the parties as to the fulfillment of reciprocal obligations. As such, the PC calls for reciprocity—the degree of agreement on the obligation of each party to provide adequate returns to the contributions of the other party (Dabos & Rousseau, 2004; Wu et al., 2006)—, and mutuality—the degree of acceptance of the parties to their respective obligations in the relationship (Coyle-Shapiro and Parzefall 2008; Dabos & Rousseau, 2004).

The forming and developing of the PC is based on individual factors—cognitive and attitudinal characteristics, ways of processing information—, and organizational factors—social messages and signals arising, for example, from the organizational culture (Dulac, 2005; Rousseau, 1995). Each PC is individual, unique, and idiosyncratic, and differences can be analyzed according to their content.

Rousseau (1995) proposes four types of PC: transactional, transitional, balanced, and relational. Transactional contracts assume specifically defined and tangible exchanges, such as the amount of pay, level of performance, and deadlines (Taylor et al., 2006; Vantilborgh et al., 2011) and limited emotional involvement of the parties (Raja et al., 2004). Relational contracts, on the other hand, present a more implicit and less tangible feature, marked by mutual trust between the parties, and are therefore structured in a markedly socioemotional and dynamic framework (Isaksson et al., 2010; Taylor et al., 2006; Vantilborgh et al., 2011), with terms oriented towards organizational loyalty, security or organizational growth (Morrison & Robinson, 1997; Raja et al., 2004).

The idiosyncratic nature of the PC admits that its transactional and relational orientations, although distinct, are not mutually exclusive (Scheel & Mohr, 2013), coexisting interactively and dynamically (Dainty et al., 2004), approached as integrating a contractual continuum (Dainty et al., 2004; Millward & Hopkins, 1998; Rousseau, 1989, 1995). The balanced PC type is much similar to the notion of shared employability, where both parties agree to share employment responsibilities concerning career development and training. Finally, the transitional PC marks the indefinition of the employment relationship between the parties due to unexpected changes in the external environment and that affect the organization.

The literature alludes to another component of PCthe ideological (e.g. Bal & Vink, 2011; Scheel & Mohr, 2013; Thompson & Bunderson, 2003; Vantilborgh et al., 2012). Bal and Vink (2011) consider that this ideological perspective allows for a broader understanding of exchange obligations beyond and above economic, socioemotional, and developmental obligations within the individual-organization relationship. Thompson and Bunderson (2003) introduce the concept of "ideological contract" arguing that there are obligations in the terms of the exchange relationship between the individual and the organization that is not part of the traditional transactional-relational framework and propose the ideological approach as an extension of the usual PC terms (see Table 1).

Thompson and Bunderson (2003, p. 574) define the ideological contract as "the credible commitments to pursue a cause or principle (not limited to self-interest) that are implicitly exchanged at the nexus of the individual-organization relationship." Several authors (e.g. Vantilborgh et al., 2011, 2012) argue that the ideological PC relies on a credible commitment to a general cause or principle as the central

Table 1 Extended typological framework of the psychological contract

	Typology	Transactional	Relational	Ideological
	Main dimension	Economic	Socioemotional	Ideological/values
Exchange content	Organizational obligations	Continuous employment, job security, fair compensation	Training, career development, promotion opportunities, long-term job security	A credible commitment to a valued social cause
	Individual obligations	Compliance with formally specified requirements	General compliance; organizational commitment and involvement; organizational citizenship behavior	Participation in the organization's mission/cause; Organizational and societal citizenship behaviors
	Main beneficiary	The self ("I")	The self and the organizational community ("we")	Society, any of its segments, or an intangible principle ("everyone")
	Affiliation logic	"The organization gives me a fair wage/day for a fair day's work"	"The organization nurtures my professional development and sense of community"	"The organization shares my passion, cause, and/or mission"
	Human nature principles	Selfish, instrumental	Collectivist, socialized	Principles-based, involved

Adapted from Thompson and Bunderson (2003) and O'Donohue and Nelson (2009)

element of the relational exchange between the individual and the employer. It follows that for Thompson and Bunderson (2003), individuals' beliefs about the ideological obligations of organizations are not established essentially based on personal rewards or exercise of rights, but rather on organizational loyalty and commitment to fulfilling the cause, mission or principle defended (organizational obligation); and on the premise that the organization provides the conditions for them to feel obliged to contribute to the achievement of ideologically structured goals (individual obligation).

In US Army combat units in Iraq, Scott et al. (2006, p. 313) refer to "devotion to the cause" as a specific performance motivator, while O'Donohue and Nelson (2007) and O'Donohue et al. (2007) studies, conducted with nurses in public Hospitals and Australian scientists, suggest that the ideological component may imply reconsidering the cognitive-perceptual definition of PC, as well as the traditional transactional-relational interpretive framework based on economic and socioemotional biases. This seems to be especially the case of professional categories guided by ethical norms.

Isaksson et al. (2010) consider that all PCs are structurally centered on transactional exchanges, to which are added, in some cases, relational obligations leading to the commitment of employees and employers beyond what is strictly necessary. Thus, they suggest that the PC can be treated as a layered model where the core is transactional obligations and the surrounding level is based on relational obligations. In accordance, Vantilborgh et al. (2012, p. 1074) highlight that this kind of "values-based" PC (Scheel & Mohr, 2013, p. 390) can be considered a third layer of the model, of great importance to explain the PC of volunteers and employees in non-profit organizations with a strong emphasis on a mission, principles or values.

As Dixon-Fowler et al. (2020, p. 2) refer, an ideological PC: "may provide employees with a sense of meaning and connection to their organization which both justifies and motivates their hard work on behalf of the employer." And this seems to be quite important due to the "decline of social-emotional elements such as job security and loyalty in modern employment relationships" (Dixon-Fowler et al., 2020, p. 2).

Based on the previous reasoning, we can expect a decline in PCs founded on relational and socioemotional terms, especially under situations of economic constraint and organizational restructuring.

3 The Military Context

The military organization is the social institution responsible for the monopoly of the "management of violence" (Heinecken, 2014; Huntington, 1957). However, since the end of the Cold War (e.g. Dandeker, 2006; Hedlund, 2013; Heinecken, 2009; Latham, 2002; Kraak et al., 2020; Manigart, 2006) the armed forces of Western countries have known changes both from an organizational-structural and professional point of view.

The unique organizational characteristics and culture of the military institution are determined by hierarchical, formal, and strongly socialized professional requirements, framed by explicit regulations and procedural norms (Siebold, 2001), typical of the organizational or traditional career typology (Baruch, 2006; Baruch & Quick, 2007, 2009; Grimland et al., 2012; Kovalenko & Mortelmans, 2014).

Commonly considered an institution—a social entity guided by a set of values, norms, and specific behaviors that transcend individual interests for the sake of a greater good, in this case, the community (e.g. Caforio, 1988; Latham, 2002; Moskos, 1977; Nuciari, 2006; Williams, 2008)—, the military organization has as its core to ensure the defense and security of the society from which it emerges, and of which it is an integral part. With the fall of the Berlin Wall (Heinecken, 2009; Nuciari, 2006) and the disintegration of the Soviet Union (Morgan, 2003), the classic concerns centered around the defense of national territory against enemy invasions and nuclear attacks, became secondary. Instead, the military assumed new roles in the defense against global threats typical of the new international security framework (Dandeker, 2006; Heinecken, 2009; Williams, 2008), which led to a rethinking of the structure, functions, and missions of the armed forces (Segal & Babin, 2000).

In the case of public organizations, such as military institutions, it may be considered common that organizational change occurs as a response to pressure from the external environment in which they operate (Farazmand, 2002). After all, military organizations are open systems with relationships of interdependence and exchange with the environment (Manigart, 2006), where processes such as organizational restructuring and downsizing are also current events in today's armed forces. Such processes (Cheng & Su, 2013; Heinecken, 2009; Nuciari, 2006) ensured that armies became smaller, highly specialized, and professionalized "with deep connections with the business sector" (Baltazar, 2005, p. 61), As such, each military action is guided by efficiency criteria in detriment of effectiveness (Snider & Watkins, 2000), generating the "post-modern military organization" (e.g. Manigart, 2006). These changes have also impacted the military profession and career perspectives (Cheng & Su, 2013; Heinecken, 2014; Kraak et al., 2020; Manigart, 2006; Snider & Watkins, 2000).

Caforio (1988) notes the existence in the Army of three categories of personnel which he refers to as "employment situations" (p. 56), two of them of long duration—officers and sergeants, and a third, varying from country to country and depending on the type of service—compulsory or other employment situation—, is somewhat similar to that of sergeants. He concludes that only the officers' category would be included within the profession.

Feld (1975) also shares the idea that the military profession is limited to the category of officers who maintain a permanent link with the organization, from recruitment to the training process that generates specific skills. As for Huntington (1957), only career officers meet the conditions that make them capable of managing violence and, as professionals, meet the characteristic attributes of "a special type and vocation" (p. 8), which would be especially linked to an ideal-type in the case of the military.

In Portugal, the term "military profession" is not common, but rather "military career". The EMFAR (2015) characterizes the military career as "a set of hierarchical positions, developed by categories, which is materialized in special cadres and to which corresponds the performance of positions and the exercise of functions differentiated among themselves" (Art. 27 of the EMFAR, 2015).

Although the military organization maintains some autonomy from societal trends (Moskos, 1977, 1986), it is not immune to the changes occurring in civil society and the increasingly globalized world. Moskos (1977, 1986) developed a two-dimensional model to characterize the trends of the military's activity: the Institutional/Occupational (I/O) model. The institutional dimension is linked to the principles, values, and norms by which the institution is guided and should govern professional conduct, and the expected return is mainly intangible, such as psychological satisfaction and social recognition. The occupational dimension is, above all, anchored in personal interests, which prevail over organizational ones, and the expected return is marked by an economic-financial order typical of the labor market. For Caforio and Nuciari (1994), the two orientations, being extremes, are not opposites—as the model presumably considers—nor mutually exclusive, but rather coexist, which allows for the generation of diverse tendencies of organizational affiliation with an institutional slant.

This model presents a certain evident correspondence with PC types, with the institutional component similar to the relational employment contract, and the occupational perspective resembling the transactional component of the contract, which concerns the attainment of strictly organizational goals.

Traditionally seen as a job for life, the military career, a clear example of an organizational career, was in the recent past associated with high job security (Baruch & Quick, 2009). This is no longer the case, as Heinecken (2009) points out, the officers in his study showed dissatisfaction related to the reduction in the number of functional positions and the blocking of career progression, with consequent frustrations in expectations and job insecurity.

In Portugal, the career of an Officer of the PPPA is statutorily framed by the EMFAR (2015), where the entry of officers from the Military Academy into the Army's Permanent Staff occurs, since 2008 with the Bologna Process, through the integrated master's degree in Military Sciences.

Depending on the length of the course, the career begins at the rank of Ensign or Lieutenant. The top of the career is reached at the rank of General.

In the Portuguese Army's career model, the military officer cannot be promoted without having remained at least the number of years foreseen for the rank he/she holds. The maximum age limits for each rank determine that if the military officer is not promoted to a higher rank by the time he/she reaches that age, he/she will have to leave active service in an "up or out" logic (Baruch & Quick, 2009, p. 271).

Over the last 30 years, there has been a common belief that, from the point of view of individual career expectations, the normal career path would lead to the rank of Colonel by most officers, and that the General rank would be accessible only to those who had most distinguished themselves throughout their military career.

This scenario has been challenged due to the closure and merger of military units and facilities and the reduction of promotion opportunities, with the consequent blocking of career progression. This, together with other changes considered as negative in the ongoing restructuring of the Portuguese Army (e.g. salary reductions and cuts in social and health benefits), has generated an increasing number of requests to leave the organization, indicating possible changes in the type of PC that the military may develop.

4 Methodology

This study was based on 42 individual in-depth interviews, in which the data treatment and analysis were performed through thematic content from a grounded theory perspective (Corbin & Strauss, 2014). We employed a theoretical sampling strategy, initially structured in quotas according to military rank.

These options were based on the organizational and historical characteristics of the context-organization under study, which discourage the use of more rigid research instruments in contexts of change (Buchanan & Bryman, 2007), as well as on the need to obtain the experiences, beliefs, motivations, and views of individuals about a specific situation (Gill et al., 2008; Rowley, 2012).

4.1 Data Collection Procedures

The 42 interviews were conducted between May 2012 and January 2015 and took place at a time of PAF restructuring. The research participants included 37 men and 5 women, aged between 25 and 55 years old and belonging to the universe of PPPA officers from the Military Academy, serving in 14 units and military establishments.

The participants represent different ranks and specialties. The interviews, with an average duration of 90 min, were conducted individually and in person, audio-recorded, and later transcribed verbatim into text. The study was granted consent by the Military organization and all the ethical concerns were taken into consideration.

4.2 Sample Sociodemographic Characteristics

Table 2 shows the socio-demographic characteristics of the participants in the study. The seniority in rank (SR) translates to the time in the respective rank and the length of service (LS) refers to the total time of permanence in the Portuguese Army.

Table 3 shows the distribution of participants according to specialties and ranks in the military hierarchy.

Table 2 Sociodemographic information of the participants

Variables							
Age n	25–30 5	31–35 9	36–40 5	41–45 8	46–50 7	51–55 8	>55 0
Sex	(3♂ + 2♀)	(7♂ + 2♀)	(4♂ + 1♀)	♂	♂	♂	
SR n	1–3 16		4–6 11		7–9 15		>9 0
LS n	5–10 3	11–15 9	16–20 5	21–25 10	26–30 7	31–35 4	>35 4

Table 3 Distribution of participants by specialties and ranks

Subcategories(*)/ ranks	Specialties										Tot
	GOC	INF	ART	CAV	ENG	TM	ADMIL	MAT	MED	PHARM	
General officer(*)	3										3
Colonel		2	1	1			1				5
Lieutenant Colonel		4	1		1	1	1				8
Major		3	2	2	1	1	1	1		1	12
Captains(*)		2	1	2	1	1	1	1	1		10
Junior officer(*)		1		1	1		1				4
Total	3	12	5	6	4	3	5	2	1	1	**42**

GOC—General Officer Corps; INF—Infantry; ART—Artillery; CAV—Cavalry; ENG—Engineering; TM—Transmissions; ADMIL—Military Administration; MAT—Material; MED—Medicine; PHARM—Pharmacy

5 Findings and Discussion

The first phase of the data interpretation process was initiated with open coding (Charmaz, 2008; Corbin & Strauss, 1990). The next stage, articulated in two phases, was performed through a process of selective coding, where, first, the generated codes (theme categories) and the text fragments that had been in the respective conceptions were systematically compared, giving meaning to the subcategories and, second, the thematic categories were generated as explanatory variables of the phenomenon under study.

5.1 Psychological Contract Contents

Following the above procedure, we identified four meanings attributed by the interviewees to the PCs contents: (1) commitment; (2) job security; (3) career progression, and (4) formal duties and rights.

The **commitment** theme seems to indicate a particular way of interpreting the individual-organization relationship, and it is assumed that the "binding ties" of the relationship exist, but that they are mostly in a one-sided direction: from the individual to the organization and not the other way around, as the following statements let us know:

> (...) Joining the Army organization is a voluntary act (...) to an organization that already exists, and if any notion of commitment was expected of me it was that the organization would give me training to follow a career, at the beginning. I think that the organization did not establish any commitment with me, I even think that it is more the other way around, I am the one who established a commitment with the organization (...).
>
> (P18, Major General, 53 years, SR=1 year)
>
> I think that ... hum (...) I understand that [the question] should be put the other way around, that is, who committed to the organization was me and not the organization to me, that is, I voluntarily entered this life and accepted the rules of the game that the organization imposed on me. And the commitment that I established with the organization is mirrored in the oath of Office, I swore to defend my country even at the risk of my own life. This was the commitment that I accepted consciously, rationally.
>
> (P22, Colonel, age 53, SR= 9 years)

This *suis generis* way of understanding the notion of commitment between the two parties of an employment relationship points to a binding univocity that finds rationality in a PC with characteristically ideological content. It is based, on the one hand, on a markedly institutional and defined image of the military universe, whose aura acts as a vector inducing perceptions of reciprocity with very specific characteristics (centered, as a relational premise, on personal and voluntary availability to serve) and, on the other hand, on the traditional individual orientation to respond to the unique missionary and elevated calling mentioned, for example, by Sørensen (1994). This univocity proved to be, understandably, common to most of the interviewees.

Job security and **career progression** are dimensions that appear perfectly defined right at the beginning of the career, which is not surprising, considering the typical organizational career characterization associated with the profession of the officer of the PPPA. These dimensions are part of the set of socioemotional terms and conditions of a relational contract type. While the first dimension remains unchanged in the current perception of almost all participants, the same is not true for career progression:

> In terms of job stability, there you go, it was a job for life. In this field, these were my expectations and I think that for now I still have my expectations fulfilled. In terms of career progression, I expected to have guaranteed reach the rank of Colonel and even, eventually, higher ranks, i.e. within the General Officer class and, guaranteed, in different timeframes than those that are happening now. (...) I would never expect to reach the age of 43 and be

(only) a Major. That is, in terms of job security the paradigm is still the same, while in terms of career progression, it is not.

(P16, Major, 43 years, SR=8 years)

One case emerges as an exception regarding job security. This exception in the relationship duration between the professional soldier and the Army may indicate a paradigm shift at the level of individual expectations of advancement and management of the career itself, especially since the opinion expressed refers to a participant who is practically at the beginning of his military career:

(...) I fit into the transition generation. When I was at the Academy it was common to say that 'if we didn't die we would become a colonel'. (...) I joined the corps at the beginning of the [economic] crisis, I no longer think that jobs are for life (...) and not everyone will reach the top positions (...) and I even think that one has to start thinking about alternative jobs after a certain age (...)

(P25, Lieutenant, 29 years, SR=3 years)

The **duties** and **rights** resulting from the answers given seem to be very much shaped by the set of duties and rights that frame the employment relationship, as set out by the legal diplomas of the EMFAR and the Military Condition Statute (Law n°. 11/89, of 01 June). When asked what they consider to be their obligations and those of the organization, they tend to refer to a set of ethical, deontological, and ideological meanings, not often alluding to terms of exchange of a transactional nature that implies, on the part of the employee-military, a restricted organizational involvement oriented towards the mere execution of the functional tasks. On the side of the employer army, the interviewees expect the guarantee of well-defined rewards depending on the work and for a determined period. Hence, the interviewees perceive their duties as follows:

The [duties are] institutional ones (...). [common also to rights]

(P14, Lieutenant Colonel, 50 years, SR=9 years)

(...) to comply and enforce orders and superior determinations, (...) the command action, the duty to optimize resources, material, and human, (...) the duties of loyalty, discipline, truth (...), to serve with zeal (...)

(P3, Lieutenant Colonel, 50 years, SR=9 years)

(...) the honor, loyalty, fellowship (...), the supreme duty to give one's life for the Homeland if necessary (...)

(P9, Major General, 53 years, SR=3 years)

And as rights:

(...) the training and learning necessary to acquire the skills to perform the duties (...), the conditions that allow me to fulfill the missions (...)

(P18, Major General, 53 years old, SR=1 year)

(...) the right of loyalty, which is simultaneously my duty in the sense that if I am loyal to the organization, it is reciprocally obliged to be loyal to me (...), and another right that I consider I have and which is, in my view, connected with what I mentioned before, which is the right to transparent information that should be given to me by the organization (...).

(P9, Major-general, 53 years, SR=3 years)

(...) I have rights such as the right to have the prestige inherent to the function that I perform and the satisfaction of my legitimate expectations (...)

(P7, Colonel, 51 years, SR=7 years)

One of the organizational duties present, implicitly or explicitly, in the narratives of all PPPA officers, is the special duty of guardianship provided for in the EMFAR—"...consists of watching over the interests of subordinates and reporting, through the hierarchical channels, problems that the military has knowledge of and that concern them" (Art. 15 of the Military Discipline Regulation; Organic Law Nº. 2/2009 of 22 July). Being, by definition, oriented towards the zeal of subordinates' interests, the guardianship concept, although generally understood as an organization duty and, as such, a right of the military professional, contains within itself the particular characteristic that is also dependent upon the organizational hierarchy occupied by the superior officer. In this way, and because it is comprised within the hierarchical structure, in which the military officer is both superior and subordinate when placed in the first of these positions, the precept is for him/her a duty and, in the second of the hierarchical positions, a right. This dual perspective makes this norm a privileged instrument for the strategic management of personnel in military organizations since it has a direct impact on areas that are peculiar to this context, such as communication, leadership, and motivation. If used appropriately, the duty of guardianship is crucial to the mutually satisfactory alignment between individual expectations and organizational goals.

Summing up, the PC terms reveal an understanding of commitment as a univocal path that entails a relational binding of the individual to the organization but not the reverse. This is an unbalanced perspective, to the disadvantage of the individual. This singular perspective between the individual "giving" and the institutional "receiving" is justified by the missionary logic of the call (Sørensen, 1994) and the specific involvement of the military condition.

Job security and career advancement are traditionally associated with the military profession and, therefore, are considered as integrating the individual-organization bond. Our results indicate that job security continues to be a contractual, formal, and psychological element that has not changed and is assumed as a permanent employment condition by all participants, while career advancement is not. There was a clear perception that the changes underway in the Portuguese Army implied profound changes in the normal evolution of vertical and horizontal promotions, namely the blocking of careers and the decreasing of opportunities for professional evolution. Bearing in mind the assumption that the management of military careers is the responsibility of the organization, the mentioned changes may generate uncertainty. This contradicts one of the primary functions of the PC, which is to reduce uncertainty (Shore & Tetrick, 1994), and may even entail a pernicious understanding of the normative nature of organizational commitment (e.g. Meyer & Allen, 1991), which seems to underlie narratives such as:

(...) I have a terrible fear of going to do the SOPC [Superior Officer Promotion Course] and not even getting promoted since I look to my left and see Majors with 8 years and Captains with 8 and 9 years. Given this, is it worth applying myself and getting a good grade in the

SOPC? (...)? Is it worthwhile for me to continue to apply and then be a Major for eight years [interrogative tone] having, of course, to consider this with the balance of family life? Will it be worth it? It's just that the answers I am given are not the ones expected of an organization that values its assets.

(P34, Captain, 33 years, SR=4 years).

Still, within the scope of job security, results suggest that the notion of employment security is associated with job continuity (i.e. quantitative experience) and is not being contemplated from a qualitative perspective. Hence, job (in)security is not considered by this military in terms of changes related to the continuity of the valued aspects of the work, such as career progression, pay, and status (Arnold & Staffelbach, 2012).

Although the duties and rights of army officers are formally established in specific legislation (EMFAR, Military Condition Statute, and Military Discipline Regulation), there is a strong reference to individual duties atypical of other professions, such as permanent availability for service, honor, discipline, fellowship, and the self-giving of one's life, traditionally associated with self-sacrifice and complete dedication to the job. As far as rights are concerned, in addition to those common in the context of a bond with a relational tendency (paternalism, pay, stability, loyalty, working conditions, and career management), typical of a traditional career, others emerge as institutional references: the right to internal information and organizational guardianship, the latter being one of the special duties of the military hierarchy.

5.2 *A Typology of the PC in the Portuguese Army*

The obligations and duties previously discussed configure the existence in this sample of only three types of PC: relational, transitional, and ideological.

The **relational** dimension of the terms of reciprocal exchange is traditionally associated with an open type of PC, essentially implicit and based on mutual trust between the parties. The PC terms, both tangible and intangible, are predominantly socioemotional, grounded in the assumption of a lifelong mutual interest involvement. Structurally based on organizational promises such as job stability, concern for employees' well-being, and the guarantee of social benefits for themselves and their families, it is natural for there to be elements of this type of contract in the employee-employer relationship in a military context:

(...) one of the big reasons for me joining the ranks was really the stability, in terms of employment. I thought, and still do, it was one of the best decisions I made (...)

(P4, Lieutenant, 31 years, SR=4 years),

(...) the access to free health care and social benefits, for me and my family, (...)

(P18, Major General, 53 years, SR=1 year).

The **transitional** tendency, not being considered a true expression of the PCbecause is not associated with mutual promises but is in a state of transition and content vagueness (Chambel, 2014; Rousseau, 2000)—appears, however, in the expressed opinions emphasizing ambiguity and uncertainty, or even, apparent loss of confidence in the organization:

> (...) until now I knew (...), I calculated in terms of years where I would be, in terms of career progression and stability, (...) before I knew where I would be and how I would be (...) I lost that part, I don't know what will happen (...)
>
> (P13, Major, 43 years, SR=6 years),
>
> (...) I feel that, from a certain point on, we became a bunch of hired guns and the feeling that I have today is of sadness, of disappointment, of ... loss, since ... having seen, having felt the organization in a certain way and today seeing that it is a monolith, a completely different thing (...)
>
> (P3, Lieutenant Colonel, 50 years, SR=9 years).

The presence of this type of contract is not surprising since the organizational changes experienced in recent years by the military and the organizational downsizing imply uncertainty (Jamil et al., 2013). Here, the emergence of the transitional PC seems to be associated, above all, with an individual disappointment resulting from the realization that not everything is being done by organizational agents—on whose shoulders the sacrosanct duty of guardianship rests. Hence, the transitional PC is manifested because the organizational responsibles are not observing the fulfillment of certain organizational obligations and, not so much, because the military is withdrawing from the relationship.

The **ideological** PC is based on exchange terms oriented towards altruism and public interest. In this way, the PC ideological beliefs are rooted in causes or principles of social interest and, therefore, particularly generated from perceptions of ethical, deontological, and human values when pursuing a valued mission, as is the case of the armed forces. This ideological tendency of the PC emerges where the defense of the homeland, the Army values, and the mission are mentioned aspects by the military interviewed:

> (...) the serving, the defending the Homeland, [is] what made me military. I'm not saying this about the initial phase when I entered, but when I decided to stay in the Army, my contract was with values, ultimately it would be [leading] my men, (...) all the oaths I took, the Oath of the Flag, the Oath of Allegiance ... I still remember, the Cadet Code.
>
> (P3, Lieutenant Colonel, 50 years old, SR=9 years),
>
> (...) I am here for [completing] our mission and not to carry a simple job, in a 'nine to five function' (...), I still believe in our institution and our mission, why we exist, and it will be sad if I stop believing in it, but when I stop believing I will not stay...
>
> (P30, Major, 35 years, SR=1 year).

A special feature of the mutuality of exchanges that underlie an ideological PC, particularly in a military organizational context, is that the individual is prepared to give more of himself than he expects to receive from the organization. The career of an officer in the PPPA includes a set of ethical and professional codes to which the military commits him/herself from the very beginning of his/her training process. One of these codes is an individual duty of special uniqueness, presumably the monopoly of the military profession, which involves the public assumption of the sacrifice of his/her life in the exercise of the profession. This is coincident with the fundamental mission of the PAF, which is: "(…) to guarantee the military defense of the Republic" (Nº. 1 of Art. 1 of LOBOFA, 2014).

This singularity is paradigmatic of a type of PC that is markedly ideological and that establishes an idiosyncratic relationship between the two parties.

According to Rousseau's model (1995, 2000), our militaries' PC corresponds to the relational and transitional types of PC. It must underlined that in our sample it was not evident the existence of a transactional contract type. This may be explained by the military condition itself, which is objectively expressed in the unique notion of individual-organization commitment of PPPA officers, and by the specifics of the career: no time limit, formally regulated, with permanent availability and a degree of involvement of the individual with the organization that presupposes the subordination of individual goals to organizational goals.

As much as Thompson and Bunderson (2003) describe, in our study, we can also observe the existence of an ideological tendency. This outcome seems natural due to the marked missionary characteristics, values, principles, and pro-social causes that guide the beliefs that shape the military context and the relationship between the individual-military and the organization-army (e.g. Kraak et al., 2020). This type of PC also implies personal sacrifice for the sake of the organization and its objectives, and the military career is understood not as a common profession, but rather as a calling (Hall & Chandler, 2005; Sørensen, 1994).

6 Conclusions

The obligations present in the employment relationship between the military and the army organization differ from the line of reciprocity between the parties involved in a common labor relationship. Hence, the military assumes that the relational bond is established, above all, from him/herself to the organization, with the prevalence of individual duties to the detriment of his/her rights. This seems similar to what was found in the study by Kraak et al. (2020) with military pilots, as they also describe their obligations as giving "200% all the time" or "social and personal life come second".

The duty of guardianship is one of the statutory duties attached to the career of a PPPA officer, with an essentially ideological content (e.g. permanent availability for service; sacrificing one's life). This duty results from the military organizational structure, which sees the officer as simultaneously performing the role of leader as well as follower within the military hierarchy.

This duality is important for the definition of human resources management policies because firstly, it has a direct impact on areas that are peculiar to this context, such as communication, leadership, and motivation, and secondly, because, if used adequately, it is crucial for the mutually satisfactory alignment between individual expectations and organizational goals.

The terms of the PC found in this study are similar to what, for instance, Kraak et al. (2020) found regarding career progression/development, training, and development, and working conditions, although in our sample the duties and rights in the employer-employee relationship are very centered in a set of duties and rights that frame the employment relationship inside the legal diplomas of the EMFAR and the Military Condition Statute, with very ethical and ideological meanings. Also, in the work of Kraak et al. (2020), job security was not an issue as it is with our military.

In the context of the typological characterization of the military PCs under study, transactional contents (Rousseau, 1995, 2000) were not evident, similarly as found in Kraak et al. (2020) study. Also highlighted by them was the supremacy of ideological tendencies (Kraak et al., 2020; Thompson & Bunderson, 2003). This picture is justified by the characteristics of a professional activity oriented towards the fulfillment of missions that sublimate values and principles, and social, collective, and patriotic interests over individual interests.

Moreover, we argue that the predominance of ideological and relational PC types is very relevant because the evidence shows that the military had few opportunities for career progression. Hence, the ideological currency of the PC is essential for motivating and leading these soldiers. Among other factors, having a relational and ideological PC allows for greater organizational commitment, less intention to leave, and greater satisfaction (e.g. Millward & Hopkins, 1998; Raja et al., 2004). All of these, are of extreme importance for the management of organizations, particularly in the PAF.

6.1 Contributions and Limitations

One of the contributions of this study stems from the specificity of being a study in a military organization, where research in the areas of people management and organizational behavior are not abundant, and where the study about the PC content is incipient, especially with PPPA officers (e.g. Dixon-Fowler et al., 2020; Kraak et al., 2020).

A second contribution impacts the theory of social exchange (Blau, 1964) that shapes the concept of PC. Our study findings demonstrate the existence of a different perspective on the commitment that is commonly used in the context of reciprocity, especially when approached in the traditional relational-transactional framework of the contract typology. Expressing an apparent and voluntarily assumed "imbalance of reciprocity" of obligations in favor of the employing party, it is worth further studying the referred notion of commitment.

Most research on PC has been conducted in private-sector organizations. Because the Portuguese Army is an eminently public institution, it is advisable, given the findings, to differentiate between public and private sectors as specific universes when studying PC terms and their consequences for human resource management.

Finally, we recommend the use of the statutory duty/right of guardianship as a privileged instrument for managing the military, since this duty is considered determinant for internal discipline and cohesion, motivation, and leadership, which is crucial to this type of organization. In addition, it contributes to an alignment between individual expectations and organizational goals. This is especially relevant because the holding of an ideological PC can be paramount when leading in difficult circumstances, such as situations that comprise organizational restructuring and the unfulfillment of PC terms. As suggested by Dixon-Fowler et al. (2020), if organizations contribute to developing ideological PC (through HRM practices) this may help to effectively recruit and retain employees due to a valued cause. That may be particularly important for the military context, which is an organizational context where, as results show, many employees may "pursue careers for ideological values, and they are willing to make personal sacrifices for the 'greater good" (Yang et al., 2022, p. 515).

Hence, the development of an ideological PC may be the answer to retain a pool of talented personnel, hence HR policies and practices must be focused not only on issues related to career development opportunities, but also on the dissemination of an organizational culture that defends values such as honor, camaraderie, loyalty, and self-sacrifice assumed by both parties in the employment relationship. In this regard, the military has to feel respected and valued by the Armed Forces.

A limitation of this study arises from the fact that one of the authors integrates the same area of activity of the participants, being hierarchically superior to most of them. Therefore, when conducting the interviews the power imbalance had to be carefully addressed. This was assured by guaranteeing the interviewee's confidentiality of the data and anonymity, as well as thoroughly explaining the research goals and the intended use of the information given by them. Although this circumstance may have worked as a facilitator of the interviewer-interviewee communication process given the thematic alignment, these may have been, at one time or another, an influential factor.

Another limitation lies in the time frame of the interviews. More than two years elapsed between the first and the last interview. This was largely due to the grounded procedures of the data processing methodology adopted. Further, this period for data collection was aggravated by the need to reconcile the participants' availability to give the interview.

A suggestion for future research points to studies that explore the notion of commitment and how is this commitment perceived when the relationship between the employer and the employee is highly power-imbalanced. Shore and Barksdale (1998) consider that a PC in which the employee feels indebted to what he or she receives from the employer is an unbalanced PC. Is it so in professions with strong ideological connotations such as the military? Ruokolainen et al. (2016) claim that research on (un)balanced psychological contracts is unclear about the obligations that determine it.

It is also important to identify the existence or not of a perception of non-compliance with the PC by the soldiers studied, and the consequences for the employer (Army)-employee (Military) relationship of this perception, especially given the context of restructuring.

References

Arnold, A., & Staffelbach, B. (2012). Perceived post-restructuring job insecurity: The impact of employees' trust in one's employer and perceived employability. *German Journal of Research in Human Resource Management, 26*(4), 307–330.
Bal, P. M., & Vink, R. (2011). Ideological currency in psychological contracts: The role of team relationships in a reciprocity perspective. *The International Journal of Human Resource Management, 22*(13), 2794–2817.
Bal, P. M., Jansen, P. G., Van Der Velde, M. E., De Lange, A. H., & Rousseau, D. M. (2010). The role of future time perspective in psychological contracts: A study among older workers. *Journal of Vocational Behavior, 76*, 474–486.
Baltazar, M. (2005). *As Forças Armadas Portuguesas. Desafios numa sociedade em mudança.* Caleidoscópio.
Baruch, Y. (2006). Career development in organizations and beyond balancing traditional and contemporary viewpoints. *Human Resource Management Review, 16*(2), 125–138.
Baruch, Y., & Quick, J. C. (2007). Understanding second careers: Lessons from a study of US Navy admirals. *Human Resource Management, 46*(4), 471–491.
Baruch, Y., & Quick, J. C. (2009). Setting sail in a new direction: Career transitions of US Navy Admirals to the civilian sector. *Personnel Review, 38*(3), 270–285.
Blau, P. M. (1964). *Exchange and power in social life.* Wiley.
Buchanan, D., & Bryman, A. (2007). Contextualizing methods choice in organizational research. *Organizational Research Methods, 10*(3), 483–501.
Caforio, G. (1988). The military profession: Theories of change. *Armed Forces and Society, 15*(1), 55–69.
Caforio, G., & Nuciari, M. (1994). The officer profession: Ideal-type. *Current Sociology, 42*(3), 33–56.
Chambel, M. J. (2014). Contrato Psicológico e Comportamentos de Cidadania Organizacional. In S. Gonçalves (Coord.), *Psicossociologia do Trabalho e das Organizações – Princípios e Práticas* (pp. 151–172). PACTOR.
Charmaz, K. (2008). Grounded theory as an emergent method. In S. N. Hesse-Biber & P. Leavy (Eds.), *Handbook of emergent methods* (pp. 155–172). The Guilford Press.
Cheng, J. W., & Su, T. W. (2013). Career plateauing and institutional/occupational intention for Taiwanese career officers. *Military Psychology, 25*(2), 156.
Conway, N., & Briner, R. B. (2005). *Understanding psychological contracts at work: A critical evaluation of theory and research.* Oxford University Press.
Corbin, J. M., & Strauss, A. (1990). Grounded theory research: Procedures, canons, and evaluative criteria. *Qualitative Sociology, 13*(1), 3–21.
Corbin, J. M., & Strauss, A. (2014). *Basics of qualitative research: Techniques and procedures for developing grounded theory* (4th ed.). Sage Publications.
Coyle-Shapiro, J. A. M., & Parzfall, M. (2008). Psychological contracts. In C. L. Cooper & J. Barling (Eds.), *The SAGE handbook of organizational behavior* (pp. 17–34). Sage Publications.
Dabos, G. E., & Rousseau, D. M. (2004). Mutuality and reciprocity in the psychological contracts of employees and employers. *Journal of Applied Psychology, 89*(1), 52.
Dainty, A. R., Raiden, A. B., & Neale, R. H. (2004). Psychological contract expectations of construction project managers. *Engineering, Construction and Architectural Management, 11*(1), 33–44.
Dandeker, C. (2006). Building flexible forces for the 21st century: Key challenges for the contemporary armed services. In G. Caforio (Ed.), *Handbook of the sociology of the military* (pp. 405–416). Springer.
De Vos, A., Buyens, D., & Schalk, R. (2003). Psychological contract development during organizational socialization: Adaptation to reality and the role of reciprocity. *Journal of Organizational Behavior, 24*(5), 537–559.

Dixon-Fowler, H., O'Leary-Kelly, A., Johnson, J., & Waite, M. (2020). Sustainability and ideology-infused psychological contracts: An organizational—And employee level perspective. *Human Resource Management Review, 30*(3), 1–13.

Dulac, T. (2005). Da formação à avaliação do contrato psicológico: Revista da literatura e perspetivas de pesquisa. In N. Delobbe, O. Herrbach, D. Lacaze & K. Mignonac (Dirs.), *Comportamento Organizacional—Vol. 1* (pp. 91–140). Instituto Piaget.

EMFAR—Estatuto dos Militares das Forças Armadas (Decreto-Lei n.º 90/2015, de 29 de maio): Diário da República n.º 104, I série, de 29 de maio de 2015.

Farazmand, A. (Ed.). (2002). *Modern organizations: Theory and practice*. Praeger Publishers.

Feld, M. D. (1975). Military professionalism and the mass army. *Armed Forces and Society, 1*(2), 191–214.

Freese, C., Schalk, R., & Croon, M. (2011). The impact of organizational changes on psychological contracts: A longitudinal study. *Personnel Review, 40*(4), 404–442.

Gill, P., Stewart, K., Treasure, E., & Chadwick, B. (2008). Methods of data collection in qualitative research: Interviews and focus groups. *British Dental Journal, 204*(6), 291–295.

Grimland, S., Vigoda-Gadot, E., & Baruch, Y. (2012). Career attitudes and success of managers: The impact of a chance event, protean, and traditional careers. *The International Journal of Human Resource Management, 23*(6), 1074–1094.

Guest, D. E. (2004). The psychology of the employment relationship: An analysis based on the psychological contract. *Applied Psychology: An International Review, 53*, 541–555.

Guest, D. E., & Conway, N. (2002). Communicating the psychological contract: An employer perspective. *Human Resource Management Journal, 12*(2), 22–38.

Hall, D. T., & Chandler, D. E. (2005). Psychological success: When the career is a calling. *Journal of Organizational Behavior, 26*(2), 155–176.

Hedlund, E. (2013). Civil–military control over the Swedish military profession: An analysis from the perspective of officer rank and officer education. *Armed Forces and Society, 39*(1), 135–157.

Heinecken, L. (2009). Discontent within the ranks? Officers' attitudes toward military employment and representation: A four-country comparative study. *Armed Forces and Society, 35*(3), 477–500.

Heinecken, L. (2014). Outsourcing public security: The unforeseen consequences for the military profession. *Armed Forces and Society, 40*(4), 625–646.

Huntington, S. P. (1957). *The soldier and the state: The theory and politics of civil-military relations* (Vol. 514). Harvard University Press.

Isaksson, K., De Cuyper, N., Oettel, C., & De Witte, H. (2010). The role of the formal employment contract in the range and fulfillment of the psychological contract: Testing a layered model. *European Journal of Work and Organizational Psychology, 19*(6), 696–716.

Jamil, A., Raja, U., & Darr, W. (2013). Psychological contract types as moderator in the breach-violation and violation-burnout relationships. *The Journal of Psychology, 147*(5), 491–515.

Janowitz, M. (1971). *The professional soldier*. Free Press.

Kovalenko, M., & Mortelmans, D. (2014). Does career type matter? Outcomes in traditional and transitional career patterns. *Journal of Vocational Behavior, 85*(2), 238–249.

Kraak, J. M., Lakshman, C., & Griep, Y. (2020). From top gun to the daily grind: Contextualizing psychological contract breach for military pilots. *European Journal of Work and Organizational Psychology, 29*(2), 243–261.

Latham, A. (2002). Warfare transformed: A Braudelian Perspective on the revolution in military affairs'. *European Journal of International Relations, 8*(2), 231–266.

Lei de Orgânica de Bases da Organização das Forças Armadas—LOBOFA (Lei Orgânica nº 6/ 2014, de 1 de setembro): Diário da República nº 167, I série, de 1 de setembro de 2014.

Manigart, P. (2006). Restructuring of the armed forces. In G. Caforio (Ed.), *Handbook of the sociology of the military* (pp. 323–344). Springer.

Meyer, J. P., & Allen, N. J. (1991). A three-component conceptualization of organizational commitment. *Human Resource Management Review, 1*, 61–89.

Millward, L. J., & Hopkins, J. J. (1998). Psychological contracts, organizational and job commitment job commitment. *Journal of Applied Social Psychology, 28*(16), 1530–1556.
Morgan, M. J. (2003). The reconstruction of culture, citizenship, and military service. *Armed Forces and Society, 29*(3), 373–391.
Morrison, E., & Robinson, S. (1997). When employees feel betrayed: A model of how psychological contract violation develops. *Academy of Management Review, 22*, 226–256.
Moskos, C. C. (1977). From institution to occupation trends in military organization. *Armed Forces and Society, 4*(1), 41–50.
Moskos, C. C. (1986). Institutional/occupational trends in armed forces: An update. *Armed Forces and Society, 12*(3), 377–382.
Nuciari, M. (2006). Models and explanations for military organization: An updated reconsideration. In G. Caforio (Ed.), *Handbook of the sociology of the military* (pp. 61–85). Springer.
O'Donohue, W., Sheehan, C., Hecker, R., & Holland, P. (2007). The psychological contract of the knowledge worker. *Journal of Knowledge Management, 11*(2), 73–82.
O'Donohue, W., & Nelson, L. (2007). Let's be professional about this: Ideology and the psychological contracts of registered nurses. *Journal of Nursing Management, 15*(5), 547–555.
O'Donohue, W., & Nelson, L. (2009). The role of ethical values in an expanded psychological contract. *Journal of Business Ethics, 90*(2), 251–263.
Raja, U., Johns, G., & Ntalianis, F. (2004). The impact of personality on psychological contracts. *Academy of Management Journal, 47*(3), 350–367.
Robinson, S. L., & Morrison, L. Y. (1995). Psychological contracts and OCB: The effect of unfulfilled obligations on civic virtue behavior. *Journal of Organizational Behavior, 16*, 289–298.
Rousseau, D. M. (1989). Psychological and implied contracts in organizations. *Employee Responsibilities and Rights Journal, 2*(2), 121–139.
Rousseau, D. M. (1995). *Psychological contracts in organizations: Understanding written and unwritten agreements*. Sage Publications.
Rousseau, D. M. (2000). *Psychological contract inventory technical report*. Carnegie Mellon University.
Rowley, J. (2012). Conducting research interviews. *Management Research Review, 35*(3/4), 260–271.
Ruokolainen, M., Mauno, S., Diehl, M. R., Tolvanen, A., Mäkikangas, A., & Kinnunen, U. (2016). Patterns of psychological contract and their relationships to employee well-being and in-role performance at work: Longitudinal evidence from university employees. *The International Journal of Human Resource Management*, 1–24.
Scheel, T. E., & Mohr, G. (2013). The third dimension: Value-oriented contents in psychological contracts. *European Journal of Work and Organizational Psychology, 22*(4), 390–407.
Scott, S., McCone, D., & Mastroianni, G. R. (2006). Psychological contracts in two U.S. combat units in Iraq: What happens when expectations and realities diverge? *Sociological Focus, 4*(39), 301–317.
Segal, D. R., & Babin, N. E. (2000). Institutional change in armed forces at the dawning of the twenty-first century. In S. R. Quah & A. Sales (Eds.), *The international handbook of sociology* (pp. 218–235). Sage Publications.
Shore, L. M., & Barksdale, K. (1998). Examining degree of balance and level of obligation in the employment relationship: A social exchange approach. *Journal of Organizational Behavior*, 731–744.
Shore, L., & Tetrick, L. (1994). The psychological contract as an explanatory framework in the employment relationship. In C. Cooper & D. Rousseau (Eds.), *Trends in organizational behavior* (pp. 91–109). Wiley.
Siebold, G. L. (2001). Core issues and theory in military sociology. *Journal of Political and Military Sociology, 29*(1), 140–159.
Snider, D. M., & Watkins, G. L. (2000). The future of army professionalism: A need for renewal and redefinition. *Parameters, 30*(3), 5–20.

Sørensen, H. (1994). New perspectives on the military profession: The I/O model and esprit de corps reevaluated. *Armed Forces and Society, 20*(4), 599–617.
Suazo, M., Martínez, P., & Sandoval, R. (2009). Creating psychological and legal contracts through human resource practices: A signaling theory perspective. *Human Resource Management Review, 19*, 154–166.
Taylor, T., Darcy, S., Hoye, R., & Cuskelly, G. (2006). Using psychological contract theory to explore issues in effective volunteer management. *Management Quarterly, 6*(2), 123–147.
Thompson, J. A., & Bunderson, J. S. (2003). Violations of principle: Ideological currency in the psychological contract. *Academy of Management Review, 28*(4), 571–586.
Van de Ven, C. (2004). The psychological contract: a big deal! In *Research and technology organization technical report*, North Atlantic Treaty Organization (NATO). http://ftp.rta.nato.int/Public/PubFullText/RTO/TR/RTO-TR-HFM-107/TR-HFM-107-03E.pdf
Vantilborgh, T., Bidee, J., Pepermans, R., Willems, J., Huybrechts, G., & Jegers, M. (2011). A new deal for NPO governance and management: Implications for volunteers using psychological contract theory. *VOLUNTAS: International Journal of Voluntary and Nonprofit Organizations, 22*(4), 639–657.
Vantilborgh, T., Bidee, J., Pepermans, R., Willems, J., Huybrechts, G., & Jegers, M. (2012). Volunteers' psychological contracts extending traditional views. *Nonprofit and Voluntary Sector Quarterly, 41*(6), 1072–1091.
Williams, J. A. (2008). The military and society beyond the postmodern era. *Orbis, 52*(2), 199–216.
Wu, J. B., Hom, P. W., Tetrick, L. E., Shore, L. M., Jia, L., Li, C., & Song, L. J. (2006). The norm of reciprocity: Scale development and validation in the Chinese context. *Management and Organization Review, 2*(3), 377–402.
Yang, Y., Brans, I., & Vantilborgh, T. (2022). Going above and beyond for your Beliefs: The effects of ideological psychological contract breach and fulfillment on pro-social rule breaking. *Employee Responsibilities & Rights Journal, 34*, 515–538.

Design of Cognition, Cognition of Design—A Cybernetics and Phenomenological Approach to Enhancing Organizational Cognition

Qeis Kamran

Abstract This conceptual paper aims to broaden the perspective of organizational studies by moving beyond sole reliance on the scientific dimension of observing the world through the method of reductio ad absurdum. Instead, it proposes the integration of second-order cybernetics and a design weltanschauung into the normative structuring of organizational embodiment within managerial frameworks. Grounded in a constructivist framework, this approach is recognized as the third pillar of human knowing. However, the application of cybernetics in the field of design remains limited. The process of reducing design to a scientific endeavor has restricted its broader implications as an area of study, diverting attention from holistic ontogenetical emerging inquiries that are essential to the essence of design. To establish the relevance of business studies, it is necessary for managers to engage in research through a second-order intervention. This entails designing organizations' avant le letter, which means designing practical artifacts within a hermeneutical constructivist loop of knowledge and becoming. Rather than focusing on misguided questions about "what is," business studies should redirect their inquiry towards the more appropriate question of "what can be." Consequently, the foundation of business administration is best understood as constructivist rather than metaphysical. This paper critically reviews influential publications in the fields of cybernetics, philosophy, design, and second-order cybernetics, and integrates their insights into a coherent framework. The authors propose that cognition, cybernetics, and design are interconnected phenomena, suggesting that business constructs potential worlds of autopoietic and homeostatic functionality as it navigates complexity, rather than solely pursuing profit. This paper sets the groundwork for the application of second-order cybernetics by positioning managers within the domain of design as creators of possible worlds (sein-können). Organizational design involves constructing appropriate and effective organizational Eigen-behavior. This approach draws inspiration

Q. Kamran (✉)
International School of Management (ISM), Dortmund, Germany
e-mail: qeis.kamran@ism.de

Department of Engineering Technology, University of Twente, Enschede, The Netherlands

Weiden Business School, Ostbayerische Technische Hochschule Amberg-Weiden, Weiden in Der Oberpfalz, Germany

C. Machado (ed.), *Building the Future with Human Resource Management*,
Management and Industrial Engineering, https://doi.org/10.1007/978-3-031-52811-8_8

from the works of Heinz von Foerster, Ernst von Glaserfeld, Stafford Beer, Herbert Simon, and Bruce Archer, establishing a strong foundation in design epistemology for implementing second-order intervention in contemporary HR management within complex organizations.

Keywords Design · Cognition · Cybernetics · Team syntegrity · Viable system model · Knowledge management · Human resources management

1 Introduction

The foundation of design as an essential pillar of human knowing was laid by Simon (1962), Archer (1968, 1976, 1979), Cross (2001, 2007), Romme (2003), and Ostman (2005), who have also delivered some solid foundations towards understanding design as modes of human and organizational engagement with the complexities of the environment. Constructing on Archer, the author treats design as a different culture from the sciences and the arts (Fig. 1) in terms of "*designerly ways of knowing thinking and acting*", which has been distinguished it in its own right (Cross, 2001, 2006; cf. Simon, 1969). While Cross (2001, 2006) and Schön (1983) took things a step further in that they advocated for going the design path of inquiry as the intellectual structure of a new field, Archer (1979) christened "*the third area*" (Cross, 2006, p. v). However, many scholars still advocate for a holistic approach as Owen (2007, p. 22) states: "... *a combination of science thinking and design thinking is better than either alone as a source of advice.*" The figure below describes Archer's original model of design, whereupon a solid foundation for human understanding and education was laid. The model puts modelling as the interpretive logic of design, while humanities are embraced via languages and mathematics via the dimensions of notation.

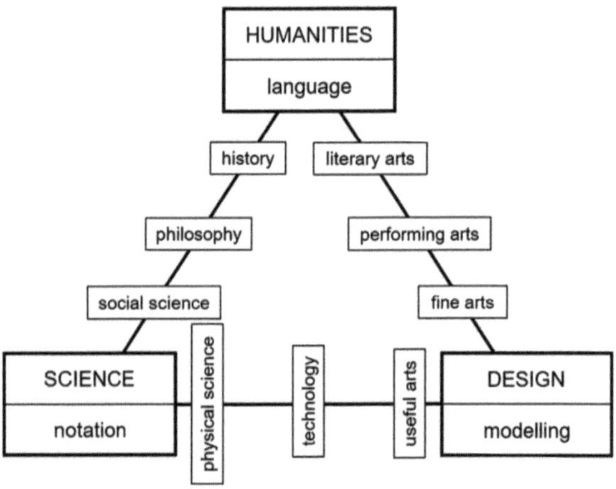

Fig. 1 Relationships between humanities, science and design. *Source* Archer (1976, 1979)

Archer's interpretation of design and putting modelling as the essence of designing corresponds with Conant Ashby's Theorem (1970) and the notion of model-based management (MBM) (cf. Schwaninger, 2010), which are the foundational laid paths of management cybernetics.

Johansson-Sköldberg et al. (2013) describe the diversity of the intellectual stream of "*design and designerly thinking* in the following manner", *which...* […]:

1. *is relation of artifacts (cf.* Simon, 1969*), applying the rationalistic epistemology*
2. *is reflexive practice (cf.* Schön, 1983*), applying the pragmatist epistemology*
3. *is problem-solving and a stating that activity (cf.* Buchanan, 1992 *based on cf.* Rittel & Webber, 1973*), applying the postmodernist epistemology*
4. *is a way of reasoning and making sense of things (*Cross, 2006, 2011*; cf.* Lawson, 2006*), applying the practical perspective/epistemology*
5. *is creation of meaning (cf.* Krippendorff, 2006*), applying the hermeneutics epistemology*
6. *is Da-sein Thinking (cf.* Kamran, 2018a, 2018b; Kamran & Neise, 2018*) ... founded within the pillars of idealism, pragmatism, phenomenology and second-order cybernetics to the contemporary modelling ... and paving the way to a more robust and holistic model of reality.*

Da-Sein Thinking has emerged as an additional stream of thought in terms of a phenomenologically "*ontic-ontological*" understanding of design, hence all realities manifested or are ex ante gleaming to which Heidegger referred to as Lichtung (cf. Heidegger, 1927; Sheehan, 2014) and perceptionally shinning (cf. Dreyfus & Kelly, 2011) within the abductive inference of human cognition of the observed phenomenon. This according to Golob and McKay (2016, p. 59) it is "*a space within which and against which entities are "disclosed" or become fully apparent*".

Archer (1968) furthermore, underpins the holistic Weltanschauung of design by uniting all the different topological distinctions of the "*design field*" such as architectural design, graphic design, and industrial design into a coherent whole.

The author proposes a new unity of design, cybernetics, and cognition as a contribution to extending the state of theory and practice of management education. As Fig. 2 illustrates this proposed unity of all three fields into one single coherent whole, as an embodiment already exists within the realm of design. Constructing within the tradition of Archer (1968), who conceived "*design*" as a verb in the broader sense of professional apperception, by stating: "*... to design is ... defined as to conceive the idea for and prepare a description of a proposed system, artefact, or aggregation of artefacts*". The authors believe that design would be much more on solid grounds if systems and cybernetics sciences were considered as the scientific pillars that unite design with human cognition. Cognition in cybernetic terms has a rich tradition and is established under the notion of second-order cybernetics. Cybernetics in the larger sense embraces the embodiment of a synergetic whole, whereupon the true nature of interdisciplinary unification of divers' scientific fields under the umbrella of design as the third way of human knowing would bear many fruits, but moreover the unity of the designer as an observer entering the domain of what is unveiled and unconcealed, is established. Cybernetics and cognition in terms of the observer entering the roam

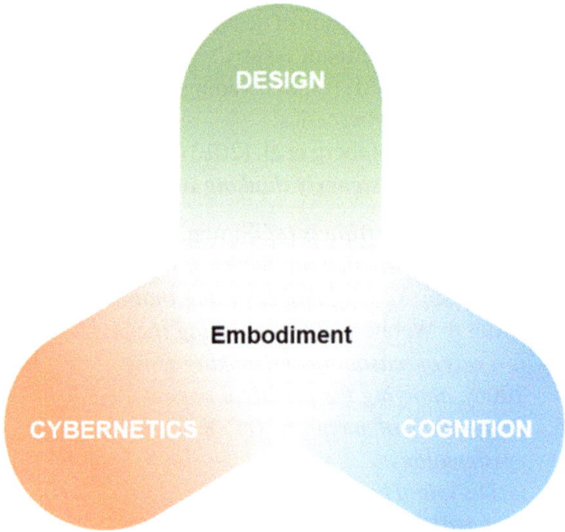

Fig. 2 Relationship of design, cybernetics and cognition. *Source* Own illustration

of what is observed, converts to cybernetics of the second-order cybernetics (cf. Foerster, 1979), which includes the observer into the domain of what is observed within the "*design search space*". This analogy is essential, hence within Simon's definition of "*design*" the notion of changing the status quo into "*preferred state*" is a "*value-laden statement*", which is decided by the observer-designer entering the roam of the design and declaring, what value is attached to the preferred state. This corresponds also to the Peircian logic of design, namely abduction, as the truth of abduction is the value-laden and workable truth (cf. Peirce, 1878) and not metaphysical.

The authors refer to the notions of "*Design, Cybernetics and Cognition*" as the nature of the same phenomenon in terms of embodiment of human knowing in navigating towards a better state within a complex environment. This dimension is described in Fig. 2, where a unified whole as the embodiment of phenomenologically constructed and cybernetically functioning realities are unified.

While the late systems-scientists and cyberneticians as Glanville (1994, 1999, 2004, 2007a, 2007b), Dubberly and Pangaro (2007), Pangaro (2016) and Fischer (2014) have made solid contributions by paving stable grounds towards design, the rapprochement from scholars of the design side has been not been mutually reciprocal. The authors regard this lack of insight as a gap within the design theory, the designerly ways of human knowing (cf. Cross, 2007) and design as the third pillar of human knowing in education (cf. Archer, 1979).

The major challenge of humans in terms of variety attenuating activity is that they must cope with complexity. Variety is the number of possible states of a system or the number of attempts within a hermeneutical "*trial and error loop*", which the designer must undertake to deliver a response to solving a problem. Design is to construct a workable reality in which possible worlds manifest themselves. Hermeneutically observed, organizational design is the embodiment of organizational ethos teleologically manifested in a circular causality by postulating functioning artifacts in an agent-based thrownness of their facticity. Effective organizations generate varieties in terms of suitable Eigen-behaviors that distinguish them from competition, which is maintaining the homeostasis of the organization as the organization navigates towards its teleological ethos and embodiment of a better world by un-concealing the hidden freedom to act towards a better state. This phenomenological notion of un-concealment originally drawing back to Heidegger (1927, 1954), is described by Popper (2012, p. vii) in the following manner: "*Every organism is constantly preoccupied with the task of solving problems. These problems arise from its own assessments of its condition and of its environment; conditions which the organism seeks to improve.*"

The notion of cognition is the apperception of designing freedom to act and achieve a state of homeostasis in terms of predicting environmental perturbations and uncertainties. Cybernetics is the bridge that connects multiple scientific fields, and it is only cybernetics that studies the spaces between sciences that usually sciences take for granted by defending the "*reductio ad absurdum*" method. What connects design and cybernetics is their mutual grounds of raison d'être of inquiries into the possible worlds by solving complex problems (cf. Beer, 2000) and wicked problems (Popper, 2012; cf. Rittel & Webber, 1973). Designers are thrown into the realm of coping with environmental complexity, where they via the "*Zuhandenheit*" of "*un-concealment*" connect the dots phenomenologically and pragmatically by delivering functioning solutions be it new or in a better way. While science investigates an observer independent reality, design and cybernetics are investigating possible realities of what could be (sein-können) (cf. Foerster, 2003; Heidegger, 1927, 1954), which cannot be unveiled without the observer entering the domain of inquiry (cf. Foerster, 2003). Design and cybernetics are observer dependent possible modes un-concealment of what could be, thus according to von Foerster: "*Objectivity is the delusion that observations could be made without an observer*" (Glasersfeld, 1996, p. 280). Wiener in his 1936 paper: "*The role of the observer*", states: "*The distinction between logic, psychology and epistemology cannot be made absolute*" (Wiener, 1936, p. 314). Simon's scientific and consultancy development was since his own cultivation of managerial mindset to solve problems was based on his training in the interdisciplinary Chicago School of Politics, where he was a protégé of Charles Merriam between 1920 and 1939s (cf. Huppatz, 2015), who applied the designerly way of knowing towards getting strategic advantages in the Cold War.

The holistic nature of design is reflected also within the original definition of Simon, where he constructs: "*Everyone designs who devises courses of action aimed at changing existing situations into preferred ones. The intellectual activity that produces material artifacts is no different fundamentally from one that prescribes remedies for a sick patient or the one to devises a new sales plan for a company or a social welfare policy a state.*" (Simon, 1962, p. 111). While Simon's seminal book has been regarded as the founding literature of the design field, another book by the father of cybernetics (cf. Wiener, 1948, 1993) Norbert Wiener (1954) called "*Invention—The care and feeding of ideas*" has been absent from mainstream design literature. The book is essential for designers of original capacity. While Wiener did not specifically call his book, a book on "*design*", because the manuscript was written in 1954, while the book was published in 1993, and the intellectual climate at that time did not observe design the way the contemporary Zeitgeist does. However, Wiener emphasized much on nature's intellectual conditions that give affordance to designers, unveiling solid novelties through their efforts via the organizations they are embedded in as "*resources integrators*" (cf. Vargo & Lusch, 2004). The challenge of what design is, is still a field of much inquiry via intellectual pursuits. Although as the field of design has matured, Page (1962 the quote is verbatim), still with justification asks: "*if one agrees that there are bodies of knowledge that have been raised here, which need further exploration—particularly a case in point would be the terminology of design—is there any point in trying to get some kind of inter-disciplinary working party going on these problems? In this question of disciplines, is there any machinery or any way of arranging for an interchange of information between specialists and people working at universities? Lastly, is there any point in making the whole thing more of a formal entity, a society, or something of that kind?*" (Lloyd, 2017, p. A1).

What scholars and practitioners need to observe is that design asks questions in pursuit of "*a how can reality*", while sciences ask the metaphysical apperceptive inquiry of "*a what is reality*". Due to the nature of the questions asked the notion of "*truth-seeking*" also changes, hence truth in terms of design pursuits "*un-concealment*" of "*possible worlds*" (sein-können) to which Heidegger refers to as "*Entbergung*" from its Greek origin of "*Alethia*", while sciences pursuit "*Wahrheit*", which cannot even be translated correctly into English. According to Foerster (1998): "*...the origin of the German word for "truth" (Wahrheit) comes from the Latin root vertias, which we in English use in terms such as verification. The English word "truth" is based on the notion of trust—possibly* via *the Proto-Germanic root "treuwaz" for belief and faithfulness*" (objective reality). While design pursuits "*Wirklichkeit*" (observer dependent reality), there is a limitation sometimes in linguistic deconstruction, hence the dual meaning in German is untranslatable into English. Constructing on Derrida's notion of "*différance*", Davis (2001, p. 14) describes: "*... is an effect of language, not a prior presence merely expressed in language. It therefore cannot be simply extracted from language and transferred*" (Kruger, 2004, p. 51).

Wiener defines invention, where semantically and as constructed above the author attributes the definition of "*design*" to it (Masani, 1996, pp. 70–71; Wiener, 1993, p. 8) in the following manner "...":

1. *Some creative intellect or intellects must conceive the idea and record it ... (Wiener, 1993, p. 8)*
2. *The material and technique needed for its execution must be available (*Masani, 1996, *p. 70)*
3. *The artisan must become philosopher, or the philosopher, artisans (ibid), in order that the idea may pass from the intellectual who conceived it to the artisan who embodies it in object (*Masani, 1996, *p. 71); and*

4. *Entrepreneurs must exist who decide that the manufacture of this object is a profitable way to make a living (ibid.).*

According to Wiener's definition, all the aspects are essential and can be regarded as the core of designing artifacts. The American laid pragmatic Western Weltanschauung of diverse dimensions such as scientific (cf. Simon, 1962, 1969), political (cf. Merriam, 1925), philosophical (Brandom, 2013; Dewey, 1934, 1938; James, 1898; cf. Peirce, 1877; Rorty, 1980), and entrepreneurial (Isaacson, 2011, 2015; Simon, 1962, 1969; cf. Wiener, 1948) have been responsible by their designerly founded modes of inquiry that brought forth the enhancement of the human condition worldwide through fostering capitalism as a model of socio-economic design to the post World War II design of the order of in West.

Constructing on Wiener's definition of design is and upon *the "ontic-ontological difference*" in Heideggerian terms, where design as an "*ontic*" mode (Zuhandenheit) of inquiry seeks the essence of the phenomenon in the possibility of its "*différance*" and being of distinction between a "*this or that determination*".

Abduction as the logic of design also underpins this mode of justification observed in defining a "*reasoning towards*" by Fann (1970, p. 4). The very nature of abduction is a second-order mode of reasoning. According to Peirce's abductive reasoning (Peirce Collected papers, p. 113): "*The abductive suggestion comes to us like a flash. It is an act of insight, although extremely fallible insight. It is true that the different elements of the hypothesis were in our minds before; but it is the idea of putting together what we had never before dreamed of putting together which flashes the new suggestion before our contemplation*". The possible entry point of observer based on cybernetics into the design search space by hermeneutically designing-forth functioning artifacts bridges phenomenology to pragmatism in an abductive sense.

2 Design and Its Philosophical Foundations from Phenomenology to Pragmatism

According to Thomas and Carroll (1979): "*Design is a type of problem solving in which the problem solver views the problem or acts as though there is some ill-definedness in the goals, initial conditions or allowable transformations*" (quoted in Cross, 2006, p. 19). Schön (1983, p. 79) substantiates this claim by stating: "*The designer shapes the situation, in accordance with his initial appreciation of it; the situation 'talks back', and he responds to the backtalk*". The pragmatic nature of the designer's work feeds back through the larger whole characterized as habits (cf. Kilpinen, 2009). To this notion, Peirce referred to as "*Knowledge is habit*" (CP 4.531; 1906), while Dewey underpinned the analogy of "*continuity*" (cf. Dewey, 1938). Thus, design thinking can be seen as a continuous process of experiences based on action and reflection, adaptation, and course correction based on circular causality of the organization's perception (cf. Foerster, 2016). Simon (1969) argued for establishing a design approach in economics and engineering and other disciplines including all the artifacts and organizations (cf. Bayazit, 2004) which has yielded some impressive results by borrowing from areas that range from architecture to NASA.

2.1 Design and Phenomenology

Design as a method of inquiry is holistic and unifies as a bridge the teleological dimension of the designer's ethos of constructing realities, thus in terms of declarative logic designer is being concerned of what ought to be (cf. Simon, 1969), to which Beer (1993) refers to via the cybernetics of organizations as "*designing freedom*". Heidegger (1927) has constructed this phenomenon of inquiry of understanding as "*sein-können*" possibilities and modes of being) in terms of making sense and understanding. Derrida coins the terms of "*l'evenir*" (to come) "... *the future, and the perhaps to open on to the coming of what comes—that is to say, necessarily in the regime of a possible whose possibilization must prevail over the impossible*" (Derrida, 2005, p. 192). Rorty's constructs upon the design of coordination and some sort of Eigen-behavior that results in a purposive action. Rorty states: "*We cannot regard truth as a goal of inquiry. The purpose of inquiry is to achieve agreement among human beings about what to do, to bring consensus on the end to be achieved and the means to be used to achieve those ends. Inquiry that does not achieve co-ordination of behavior is not inquiry but simply wordplay*" (Rorty, 1999, p. xxv). To Rorty, the divisions between finding and making, discovery and invention, objective and subjective are distinctions that ought to be eliminated (cf. Rorty, 1999). What is there may exist in a way or so, but description needs observation, which cannot be interpreted without the human mind, to which (Merlau-Ponty, 1974, p. 201) also refers as: "*We can only think the world because we have already experienced it*". The recursive nature of design treats the mind and body problem not purely from the Cartesian heritage of duality but moreover from a ubiquitous relation between cognition and action of unity of both, thus being a whole.

An extended bridge for design in addition to the pragmatic Weltanschauung can be based on phenomenology, thus this notion sets out to construct the understanding of design as stipulated and substantiated in the context of its being the science of the artificial (cf. Simon, 1969) and the third way of inquiry (cf. Archer, 1979; Cross, 2006). Phenomenology explains that knowledge is achieved by the interactions between designer and objects and can thus be seen as a countermovement against the positivistic natural sciences view, which believes that there is knowledge independent of the human perception. Therefore, phenomenological research is considered subjective, inductive, and dynamic (cf. Reiners, 2012). Designers are thus in a well justified sense constructivists of realities of what can be, "*sein-können*" *and* "*l'avenir*" (Derrida, 2005; Heidegger, 1927; cf. Isaacson, 2011; Simon, 1969), that require a recursive apperception of acting and doing based on the dimension of experience, meaning, and doing in a circular causality. According to Davis (2001):

"*...the meaning of things is not inherent in objects, but is actually located in the individual's inner life... The researcher's task is to understand reality as it is, actively and consciously created by subjects, not as a pure entity that exists 'out there*'" (Swingewood, 1984, p. 5). Hence, design as a field of holistic discourse requires a phenomenological approach of meaning and experience, of intuition and active thinking, and of means and ends analyses. It is within the dimension of high-capacity observations of having a value and adjacent consequences to it, that the phenomenon of design (Ereignis) emerges and enframes itself (cf. Heidegger, 1954).

2.2 Pragmatism the Philosophical Logic of Design

The birthplace of pragmatism can be traced to August 26th 1898, when William James held his lecture "*Philosophical Conceptions and Practical Results*" at the University of Berkeley (cf. Bernstein, 2006). In this address, James gave a tribute to Peirce for being the founder of pragmatism and expanded on the latter's Popular Science Monthly publication "*How to make our ideas clear*", by stating: "*Beliefs, in short, are really rules for action; and the whole function of thinking is but one step in the production of habits of action. If there were any part of a thought that made no difference in the thought's practical consequences, then that part would be no proper element of the thought's significance*" (James, 1898, p. 290; Peirce, 1878, p. 290). While the notion of pragma, which means behavior as christened by Peirce is in its essence action-oriented, there are still some differences between the interpretations of the various classic pragmatists. Peirce's pragmatism is more based on a logical principle of deduction, induction, and abduction. James by contrast, puts the notions of experience, consciousness, and emotion more into the context of pragmatism, thus for him, consciousness and feelings were more than what could be expressed by language. Dewey on the other hand saw experience rather as the means and ends of inquiry (cf. Rylander, 2012), constructing on the aesthetic experience of art as language and continuity in a natural environment as the mantra of concentration of pragmatism (cf. Dewey, 1934, 1938). The notion of pragmatism is very broad and difficult for any author to avoid lacuna, however from the design perspective one can align the contributions of all the essential pragmatists through a Darwinian analogy of interaction of the subject with the environment; thus whatever works, works in the sense of an aesthetic unity of experience and meaning by the ubiquitous interaction of practice (pragma) designed as a functioning (cybernetic) continuity. This is observed by Simon (1969), who states: "*Biologists are familiar with this property of adaptive systems under the label of homeostasis. It is an important property of most good designs, whether biological or artifactual*" (Simon, 1969, p. 8). Schwaninger (2006) underpins this notion "*Design for viable organizations*" by substantiating that design defines the diagnostic power of organizations. According to Brandom: "*For the pragmatists, experience is not an input to the learning process. It just is learning: the process of perception and performance, followed by perception and assessment of the results of the performance...*" (Brandom, 2013, p. 9). Designers therefore first design the organizations, which based on the integrity of the structural dynamics and interaction with the environment generate Eigen-behaviors, thus producing adequate varieties to equate with the perturbations of the environment in terms of requisite varieties (cf. Ashby, 1958) and creating the conditions of homeostasis and autopoiesis as the foundation of designing regulation.

2.3 Design Thinking

Design, which has come to be popular in management sciences as "*Design Thinking*", is rooted within the tradition of pragmatism (cf. James, 1898). From the launch of research-based journals in design thinking such as Design Studies (1979), Design Issues (1984), Design History (1988), Research in Engineering Design (1989), and Languages of Design in 1992, the term has been much popularized (cf. Cross, 1999). According to Cross (2007, p. 55), "*design practice does indeed have its own strong and appropriate intellectual culture... we must avoid swamping our design research with different cultures imported either from the sciences or the arts* (Dalsgaard, 2014, p. 144). Generally, research in design thinking is based on the below dimensions (cf. Bayazit, 2004):

1. *Physical embodiment of artifacts—performance and function*
2. *Human activity—cognition and action based on ubiquitous recursion*
3. *Purposeful activity—how an artificial thing appears and what it means*
4. *Embodiment of configurations—scientific practice*
5. *Systematic search and acquisition of knowledge related to designing*

The authors claim that Heidegger also constructs a bridge for design research in terms of "*Ge-stell*" (Enframing). He states: "*Enframing means the gathering together of that setting-upon which set upon man, i.e., challenges him forth, to reveal the real, in the mode of ordering, as standing-reserve*" (Heidegger, 1977, p. 10).

2.4 Radical Constructivism

Design is the construction of realities that are carried-out by the designer as a ubiquitous process of acting and thinking leading to constructing of even more favorable realities. One of these notions that can be discussed within the dimensions of design as the theory of "*radical constructivism*" (cf. Glasersfeld, 1995). Von Glasersfeld derives his theoretical framework from Piaget, however, there have been many pioneers, who have played an essential role in the development of the theory as Watzlawick (1984), Glaserfeld (1984), and Foerster (1981). Accordingly, the theory of radical constructivism is based upon, as Glasersfeld (1995, p. 14) states: "... *the "real" world, [and] manifests itself exclusively there where our constructions break down*". The term "*radical*" can be established as: "... *radical difference concerns the relation of knowledge and reality. Whereas in the traditional view of epistemology, as well as of cognitive psychology, that relation is always seen as a more or less picture-like (iconic) correspondence or match, radical constructivism sees it as an adaptation in the functional sense*" (Glaserfeld, 1984, p. 3). This closure into functioning again substantiates the authors' claim that epistemologically design feeds back into itself and is therefore circular. Where the designers observe or create realities, which create additional functioning realities in the Popperian refutation and falsification sense of discarding, what does not work (cf. Popper, 2012) and does not have a cash-value (cf. James, 1898). According to von Foerster, this notion was stated as: "*... to navigate is to construct*" (hvF, archives). The design world functions in multiple worlds of perceptions, interpretations, and constructions. The ability of a good designer is not to hold the one absolute reality but the development of the ability to hold and juggle with multiple complementary and competing realities by constructing them into a coherent whole. Therefore, design is the construct of an action that produces additional actions, which means that design is used as an end or an instrument towards another end in a hermeneutical sense. Glaserfeld (1984, p. 3) indicates: "*... all of us—scientists, philosophers, laymen, school children, animals, indeed any kind of living organism—face our environment as the burglar faces a lock that he has to unlock in order to get at the loot*". According to Blumrich (1979, p. 1551) "*The ability to design well is an obligation and an opportunity...*". The construction of new forms requires design, hence from the form of habits construed as the inheritances of cultural forms in terms of habitus (cf. Bourdieu, 1984), the quality of observing of future forms arises.

3 Cybernetics, a Necessary Synergy for Design and Management

In essence, cybernetics or the science of control in the animal and the machine (cf. Wiener, 1948) is about understanding and managing complexity. Wiener's work defined a new science that started the information age and various branches of informatics, systems sciences and complexity sciences. Cybernetics explains regulation in all complex systems. To understand any complex system from economies, social systems as organizations, to artifacts, which are applied within a controlled environment to robots that engage with a complex environment in terms of movement, it is necessary to understand, how they are controlled. Control is the dynamic of structure and the essence of embodiment. The notion of embodiment is better described via the principle of recursion.

3.1 The Principle of Recursion

Recursion is one of the foundational principles of management in animal, machine, and productive social systems. According to Beer: "*that every viable system contains and is contained in a viable system... If we decide to define a social system by recursion, we shall find that every viable system contains a viable system*" (Beer, 1979, p. 287). Viability in terms of systemic respectively cybernetics sciences means being "*... able to maintain a separate existence*" (Beer, 1995, p. 113). According to Beer (2002), the laws of viability in complex organisms are related to the dynamic structure regulating the adaptive connectivity of the organism's parts rather than to energy impelling the organism.

Embodiment is the triadic unity of cognition, body, and environment (Kamran, 2013; Pfeifer & Bongard, 2007; cf. Pfeifer & Scheier, 2001). A good example to demonstrate embodiment is illustrated in Fig. 3. The specific conditions, which are responsible for high performance of firms are described in terms of:

1. *the innovative orientation of the embedded firms in terms of its strategic Weltanschauung—Brain of the firm (cf.* Beer, 1972, 1981*)*
2. *the common purposive innovational infrastructure of the firm—Body—"structure is strategy" (cf.* Kamran, 2013*)*
3. *the viability of the cluster's-specific conditions, environment (cf.* Porter, 2008*), whereby these features are embedded in each other* via *the principles of recursion. Thus, the very viability of the cluster is maintained. This viability of the region is the outcome of the symmetry of the principle of recursion that deliver the necessary conditions for growth and prosperity.*

In cybernetics, the notion of control is not imposed on the system but it is the essential foundation to making sense in terms of a variety of attenuating tools to cope with the thrownness in the reality of being in nature. The raison d'être of design in nature is the symmetry of beauty meeting the ultimate functioning realty of invention and innovation. Thus, so too do the natural sciences, as Durac states: "*Physical Laws should have mathematical beauty*" Paul Durac (1955 quoted from McCubbin, 2004).

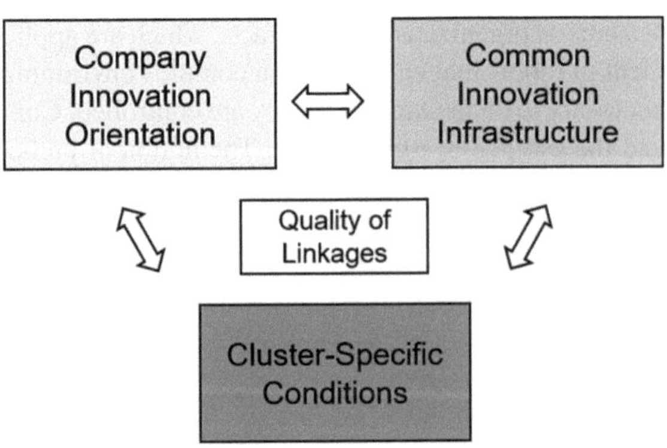

Fig. 3 Innovation in a cluster. *Source* Porter (2008)

3.2 *The Principle of Self-organization*

All natural-order and evolutionary phenomena underlie a simple but fundamental law, namely "*self-organization*". Embarking on studying of complex systems presupposes the apperception of this phenomenon. Pioneering research in social sciences has been done by Malik (1986, 2015), Ulrich and Probst (1984), Ashby (1962), Beer (1959, 1972, 1984, 1985, 1993), who put self-organization at the core of the bionics inspired "*Viable System Model*" (VSM). The most visible demonstration of self-organization as a natural law, is the self-management of the ecosystem to self-regulation of the national and global economy,[1] to a flock of birds and sea gulls self-organizing themselves into a giant whole, while each individual bird's next action and move in the flock or school of fish is not limited and hindered by any calculations and predictions (cf. Mitchell, 2009). Thus, they are still able to manage one of the most vital tasks essential to their survival (cf. Malik, 2007). Imagining giving one of the birds in the flock the position to be the leader of the group, one is sure to experience that their journey might turn not very successful. By understanding and applying self-organization as the major pillar of control "*from parts separated to parts joint*" (cf. Ashby, 1962), this collective output of the respective systems' behavior is called Eigen-behavior. The notion of "*Eigen-behavior*" (cf. Foerster, 2003; Varela, 1984) of complex systems is essential to highlight here, to which Beer refers to ascending from the recursion of the interacting sub-systems of the whole, by describing it as: "*the purpose of the system is what it does...*" (Beer, 2002, p. 218), and von Foerster coined the term and "*...equated the ability of an organization to classify its environment with the notion of Eigenbehavior*" (Rocha, 1996, p. 1). All complex systems are self-organizing, viable, and autonomous systems. As von Hayek defines: "*... the only possibility of transcending the capacity of individual minds is to rely on those super-personal 'self-organizing' forces which create spontaneous order*" (Hayek & Hamowy, 2011, p. 54). Malik and Probst underpin: "*As managers, we have to ... learn to be what we really are: not doers and commanders, but catalysts and cultivators of a self-organizing system in an evolving context*" (Malik & Probst, 1984, p. 118). By controlling complex systems, designers have come to understand that another approach and context needs to be applied departing from the traditional scientific "reduction ad absurdum" approach, thus it is holism that defines design. Design is the functioning of an artifact by maintaining the interrelated symmetry of the value projected towards an artifact and the performance delivered by maintaining the homeostasis of means and ends in the phenomenology of value-laden semantics of artifacts. Paczuski and Bak (1999, p. 1) state: "*Unless one is willing to invoke an organizing agent of some sort, all these phenomena must be self-organized*". Evolution is nothing else but a history of self-organization (cf. Kamran, 2013). Emergence welcomes change and takes its forms by adaptation in an order that it actually can survive. The quest for survival is nature's design of the ultimate doctrine. Order and chaos are just as much a natural phenomenon as it is the shape of "*fractals*". Knowing that a solution can be complex and simple simultaneously depends on

[1] Adam Smith's "The invisible hand" is a prime example of self-organization of the economy.

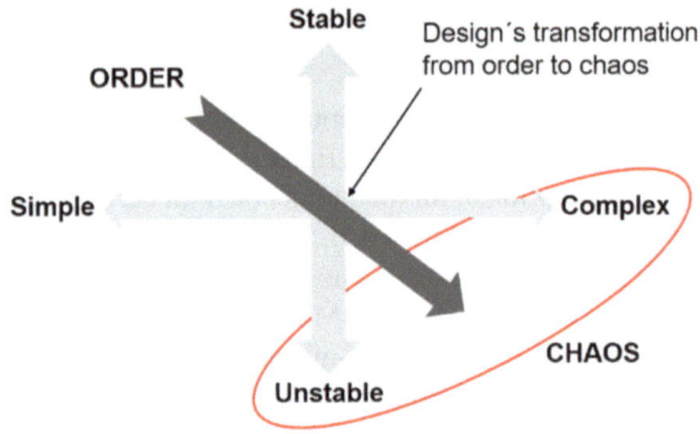

Fig. 4 The transformation of strategy from stable environments to environments of high velocity and turbulence. *Source* Own illustration

the mental model the designer is running intellectually by his cultivated cognitive apperception (Fig. 4).

Design as the third way of knowing in education is the holistic cultivation of organizations understanding and how to design artifacts as organizations to cope with the emerging shift from order to chaos of the global turbulent environment.

3.3 Ashby's Law of Requisite Variety

Ashby is the founding doyen and pioneer of cybernetics and systems theory. His major contributions are the "*Law of Requisite Variety*" (cf. Ashby, 1952, 1958) applied by Beer to construct the field of management cybernetics (cf. Beer, 1972, 1981, 2000). The principles, which are essential for management as self-organization, design, control, functionality, and governance, are the very principles of the science of cybernetics. According to the "*Law of Requisite Variety*"—"*the variety in the control system must be equal to or larger than the variety of the perturbations in order to maintain stability*" (cf. Ashby, 1958), and thus, "*only variety can absorb variety*" (Beer, 1985, p. 26). Figure 5 describes Ashby's Law and displays how designers can maintain control via solid management of operations by creating equilibrium between the organization and the environment in terms of variety engineering.

Another vital contribution of Ashby's work is the "*Conant and Ashby's Theorem*", which defines: "*Every good regulator of a system must be a model of the system,*" (Conant & Ashby, 1970, p. 1). Hence, they underpin: "*The first effect of this theorem is to change the status of model-making from optional to compulsory.*" Conant and Ashby argue that model-making has been so far generally viewed as a nicety for regulating complex dynamic systems; however, they indicate the proven necessity of a "sufficiently similar model of control" for successful regulation as the necessary pre-condition of control. Thus, to control is to design the model of the "*to be controlled situation*". The construction of such a model may have been realized explicitly or through improvement of the regulator. This suggests that based on the knowledge that "*r*egulators must model what they regulate", which is also defined by Archer

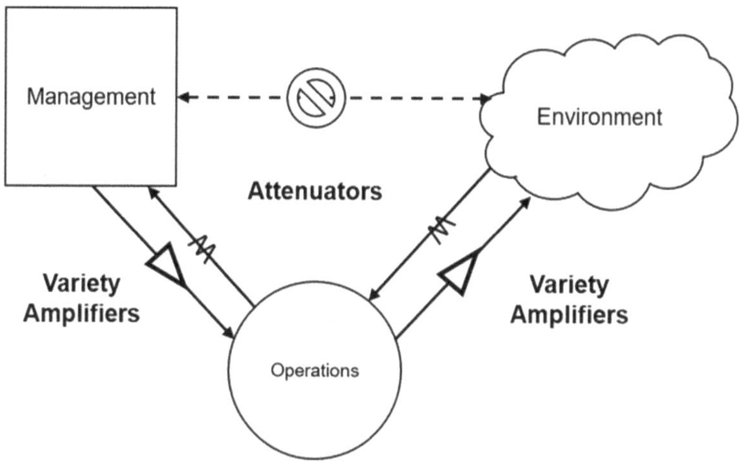

Fig. 5 Ashby's law. *Source* Ashby (1958), Beer (1979)

(1976, 1979), hence he illustrated modelling the language of design. Conant and Ashby state: "*There can no longer be a question about whether the brain models its environment: it must.*" (Conant & Ashby, 1970, p. 97). Schwaninger (2010, pp. 1419–20) also claims that "*…high-quality models can make a powerful contribution*", *hence designers are required to finalize the most essential task of this era, therefore robust designs are a necessity.*"

3.4 Variety or the Number of States in System

The designing of "*variety*" is fundamental to the management of all types of organizations. It is upon this very dimension that the whole capability of management as the embodiment of control relies. Ashby's Law is based on balancing of variety between organization and the environment. Variety engineering is the term whereby the highest abstraction of management can be described. Designers and managers coping with the environment may be able to isolate their responsibilities from the world outside, but they cannot avoid the outside forces and challenges affecting their organization. While the environment perturbs the organization by disturbances from changing dynamics in competition, consumer behavior, and demographics to changes in regulatory affairs, disruptions in innovations from products to processes to business models, to vital challenges in political dimension of the globalized world, thus, it is within the abilities of the designer to construct a variety attenuating capacity and a control system that is can cope with the situation not yet confronted, as a form of organizational intelligence. That is the reason that lean, flat, or bureaucratic systems are not capable to absorb the varieties of a dynamic and complex environment. The

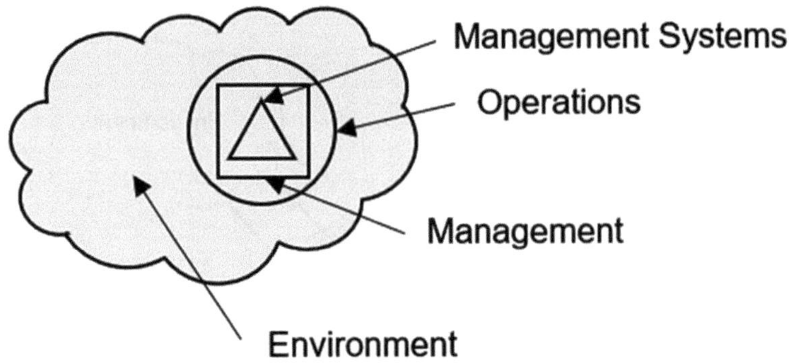

Fig. 6 A firm or organization system embedded in its environment. *Source* Kamran (2019) after Christopher (2007)

ability to model variety attenuation is essential to design a ubiquitous regulatory control system (Fig. 6).

Figure 7 describes this very relation, whereby the disturbance of the environment is transmitted to the management via the organization, which can also be described as processes or operations. The managerial function absorbs the disturbance or situation and maintains the organizational control via the managerial systems, which are highlighted below.

According to Ashby (1960), who defines this very notion: "*Two systems of continuous variables (that we called 'environment' and 'reacting part') interact so that a primary feedback (through complex sensory and motor channels) exists between them*" (Ashby, 1960, p. 98). Maturana and Varela (1980, p. 1) state: "*Organizations are adapted to their environment and it has appeared adequate to say of them that their organization represents the 'environment' in which they live and that through evolution they accumulated information about it, coded in their nervous systems*". Thus, this gathering of information and what is coded within the organization's (organisms) intelligence system (nervous system) is its cognitive activity of self-reference and thus of autopoiesis (cf. Varela et al., 1974). Von Foerster's Theorem

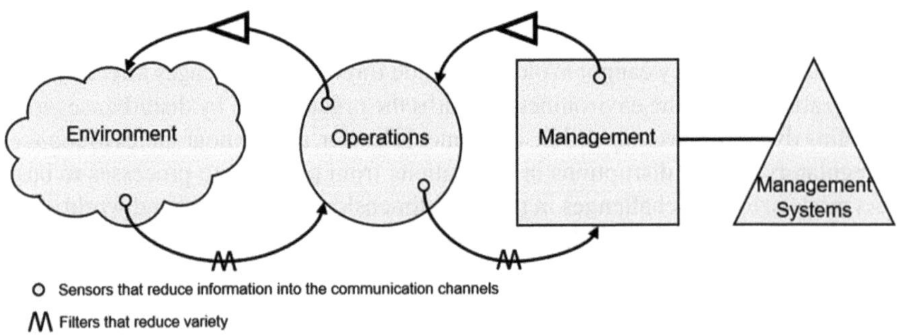

Fig. 7 The system of the firm. *Source* Kamran (2019) after Christopher (2007)

number one describes: "*always act so to advance the number of your choices*" (Foerster, 2003, p. 6).

4 Viable System Model (VSM)

Management cybernetics developed by Stafford Beer is the field of scientific management based on the aforementioned holistic view and the application of cybernetic principles originally developed by Wiener (1948), Ashby (1952, 1958, 1962), and Pask (1961). The VSM, designed and operationalized for managing organizations, is one of the most advanced, un-falsified, and valid models of organizational structures (cf. Kamran, 2013). Derived by Beer (1972, 1985) from the human brain and nervous system, the VSM is based on characteristics of living organizations.

4.1 VSM's System 1 (Operations)

At the core of management, cybernetics underlies the notion of functioning of an organization within the embedded environment. These in the living system are the organs of the body, the muscles, skin, ears, hands, feet, etc. whereby the operative functions of the body are represented by coping with the outside world. Transforming this pattern into the business, one speaks of the sales and the product service of the organization dealing with the marketplace. The best way to explain what system 1 is also by Fig. 8, whereby the authors illustrate the responsibilities of the system 1. The VSM can be designed based on the needs of the organization. Depending on the nature of organization the system 1's units can be distinguished and integrated. Figure 8 describes system 1 as the general operations unit based on the extended version of Porter's Value Chain Model (cf. Porter, 1985). This also correlates to the notion of the recursive nature of organization sub-units.

The system 1 as the operational sub-system of the VSM can be extended to as many units and divisions as there is the requirement for it. Hence, to reduce the complexity the recursion principle is fundamental to VSM, thus, all these units can be designed in a way that their control function is self-organized, hence reporting to the higher sub-systems in recursive loop.

Figure 9 displays VSM's system 1 interacting with the higher systems based on diverse channels, while the half circle figure on the left shows system 1 and the triangle shaped part is the VSM's system 2 connected to the box-shaped part displaying system 3. These additional systems will be explained in the below part.

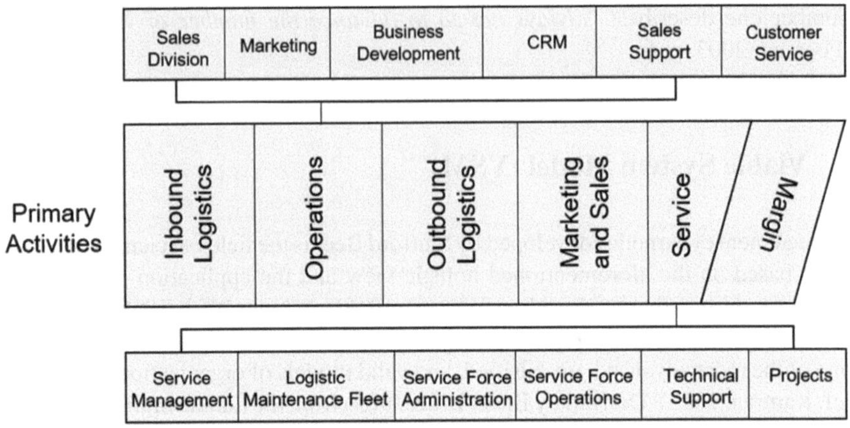

Fig. 8 Business activities according to system 1. *Source* Porter (1985)

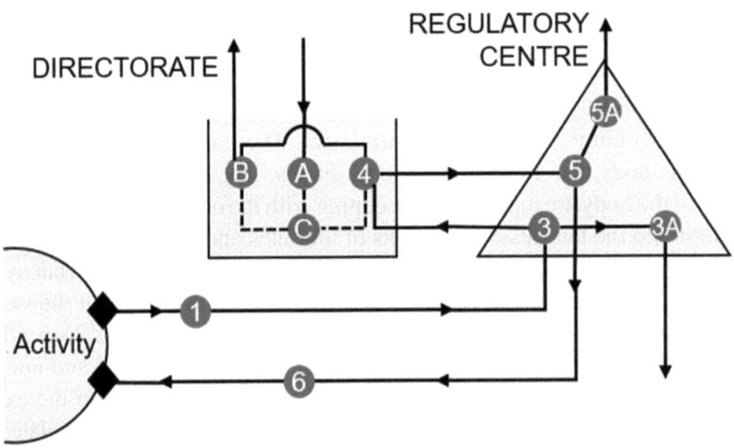

Fig. 9 VSM's system 1. *Source* Beer (1972, 1979, 1985)

4.2 VSM's System 2 (Coordination/Supporting)

System 2 as the diagrammatically described above is the tall and thin rectangular box drawn around the column of boxes, which are the system 1 (Beer, 1972, p. 172). The main function of system 2 is the coordination, connection, and the interlinking between the divisions' regulatory centers with the corporate regulatory centers. Beer states: "*So it would be correct, and even helpful, to think of system two as an elaborate interface between system three and system one. It partakes of both*". The most essential four functions of the system 2 are:

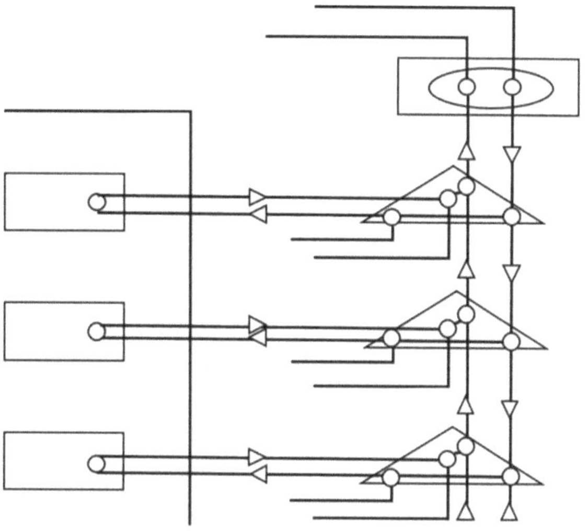

Fig. 10 VSM's system 2. *Source* Kamran (2019) after Beer (1972, 1981, 1985)

1. Coordination of the actions of the system one and the interrelation among them. It serves as the fundamental hemostat to make sure that the matters are running in operations or system one in accordance with the overall policy.
2. Control of budgetary affairs and transfer of information and collaborations with system 3 (Fig. 10).
3. Transducing the flow of information from the system 1's higher-level management (Directorate) and transmitting this information to the systems 3, 4, and 5.
4. The communication and the control function of compliance of the higher level of management by the operations (system 1).

4.3 VSM's System 3 (Direction of Internal Operations and Now)

The function of the VSM's system 3 is to manage the internal affairs and stability of the whole VSM system. This function must be considered the management and the maintaining of the homeostasis of the whole organizational internal body beyond the homeostasis of the units below within their environments and the avoidance of the internal oscillations between them. Since systems 1 and 2 are coping with situational perturbations at the lower recursion levels and keeping the local disturbances within their environments under control, it is necessary for system three to act viable as a whole and therefore navigate according to the larger internal challenges the system faces as a directory unit for the internal environment and another point that needs to underpinned here, which is the time frame of now. The primary function of system 3 is managing the internal homeostasis of the VSM in present time. System 3 enables

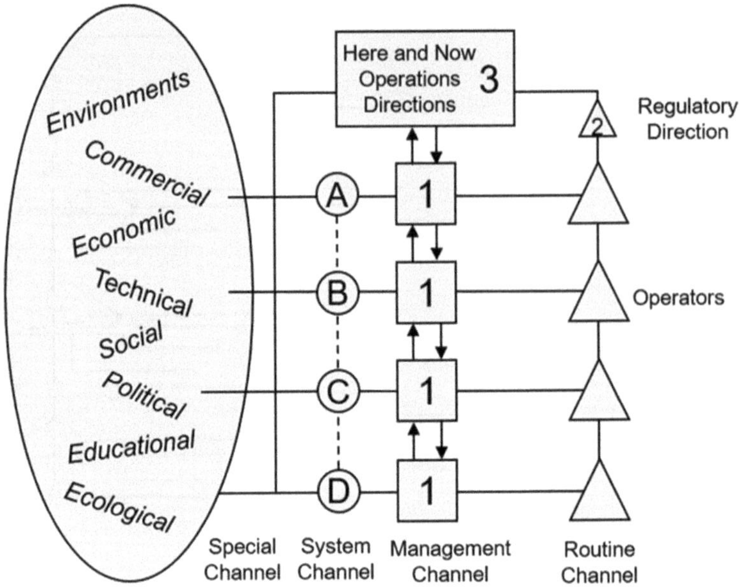

Fig. 11 VSM's system 3. *Source* Kamran (2019) after Beer (1972, 1981, 1985)

the organization to act as a whole and from a holistic perspective; it is the managerial coordination's authority for the whole organization and is considered to be the interface between the autonomous units and the highest authority (Malik, 1986, p. 119). System 3 is similar to the function of the COO of the firm, hence with a difference that it acts according to Ashby's Law by managing the inside world of the organization and the time zone "now" (Fig. 11).

4.4 VSM's System 4 (Strategic Direction/Outside and Future)

System 4 resembles all the functions and organizational parts that report to the CEO, thus, in the VSM "*the-reporting-to*" is not reconstructed but moreover, it displays the functions and communication links. In the VSM hierarchy may not be understood in terms of command and control but moreover in terms of the sum of functions that resemble a viable larger whole, hence, every function is essential and is based on ensuring survivability not displaying importance. Systems within the VSM rely on their own self-organizing, self-controlling, and self-steering autopoietic capabilities. In order for the system four to be effective and to do its duties efficiently and intelligently by identifying the value potentials, it ought to be based on "*Conant and Ashby's Theorem*". The system 4 must be able to model a favorable future, identify and design the path towards the value potentials, whereby the organization's future

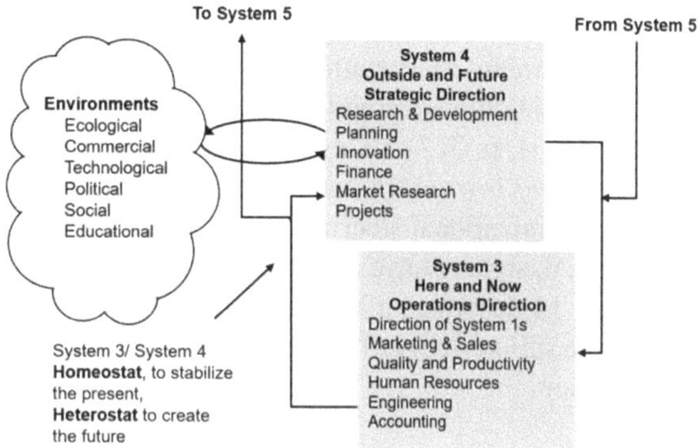

Fig. 12 VSM's system 4. *Source* Kamran (2019) after Christopher (2007)

direction needs to be constructed. According to the theorem, the matter of high-quality operations is not an operations excellence by some measures, but moreover, it is an organizational necessity and designing of its freedom. The value of every cybernetic system's model, system 4 included, regardless of its simplicity resembles the dynamic structure and characteristics the organization faces; thus, it facilitates the examination of corporate plans on an indefinite time base, which invalidates so many static models of the corporate economy (cf. Beer, 1972). Figure 12 displays System 4.

4.5 VSM's System 5 (Strategic Foresight and Executive Direction)

System 5 is the thinking part of the VSM and the direction of the organization. Foresight according to Drucker is one of the common traits he has found in successful corporations and high achieving entrepreneurial endeavors, namely the commitment to approach and integrate innovation as a "*systematic practice*" within the organization's activities. Beer has not only given the foresight activity a systemic character but moreover, he has integrated foresight into the very structure of the organization. Thus, the VSM's system 5 is acting based on foresight not only in a systematic manner but moreover it is navigating the organization based on real-time and ubiquitous connected whole within the total environment of the firm. Most vital functions of the system 5 are to give the overall leadership insights and guidance to maintain the homeostatic stability between the management of the system 3 (today and now) and system 4 (future and then) strategies. System 5 resolves as the firm's overall leadership function and manages the possible conflicts, which may arise between both realities (today and now and future and then) and organizational functions of system 3 and system 4. Maturana refers to this notion as "*autopoiesis*". "*An autopoietic system can be described as a random dynamical system, which is defined only within its organized autopoietic domain*" (Bourgine & Stewart, 2004, p. 327). According to

Maturana and Varela (1980): "*All living systems are autopoietic systems....*", thus…" *All living systems are cognitive systems*" (Fig. 13).

An organization is a cognitive and autopoietic system. According to Bourgine and Stewart (2004, p. 327): "*A system is cognitive if and only if sensory inputs serve to trigger actions in a specific way, so as to satisfy a viability constraint*". The diagnostic and the organizational steering power of the VSM is by far the most stable form of regulation designers have applied to the cybernetics of organizations.

Figure 14 describes a VSM based recursive multi-unit organizational structure, where one sub-system or business unit is contained within another to create a larger organization.

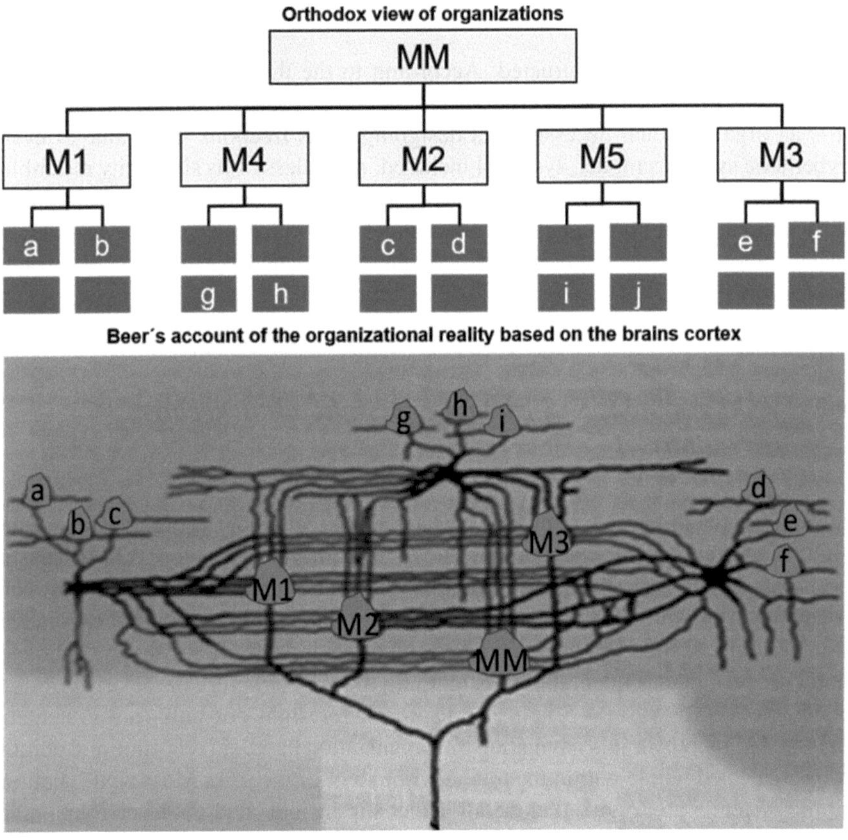

Fig. 13 Organizational charts versus how organizations really interact. *Source* Kamran (2019) after Beer (1972)

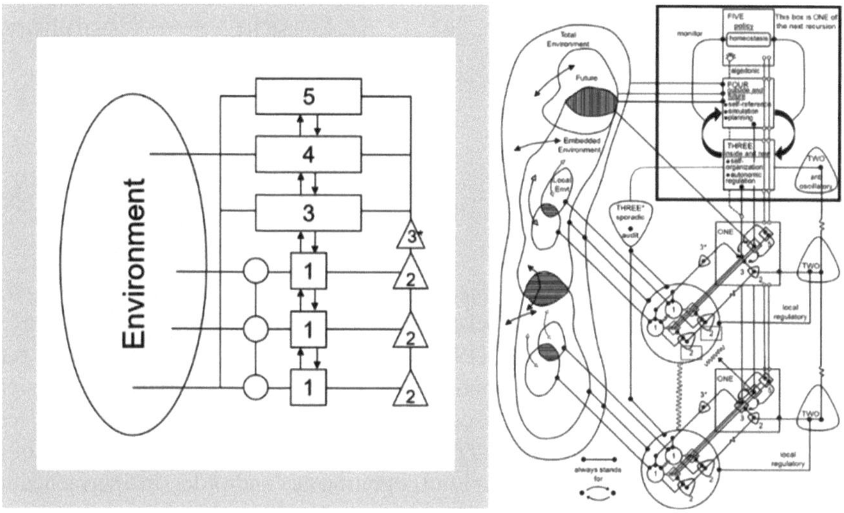

Fig. 14 Two versions of Beer's VSM. *Source* Kamran (2013, 2019) after Beer (2000)

5 Second-Order Cybernetics

Second-order cybernetics can be traced back to the conference in Philadelphia "*Communication and Control in Society*" in 1974 (cf. Krippendorff, 1996), where Heinz von Foerster established the notion of "*observing systems*" as cybernetics of second-order or "*cybernetics of cybernetics*" based on Wiener's (1948) work "*Cybernetics*". By taking the second-order Weltanschauung, the designer factor enters the observation and thus "*the shift from a first-order to a second-order cybernetics signaled a shift in scientific attitude toward reality, from privileging the perspectives of detached observers, spectators or engineers of a world outside of themselves to acknowledging our own participation in the world we observe and construct as its constituents*" (Krippendorff, 1996, p. 311). Based on the Piaget's constructivist approach and "*Radical Constructivism*" of Glasersfeld (1995), Foerster (1992), Maturana (1980) "*The Biology of Cognition*" and Maturana and Varela (1980) "*Autopoiesis and Cognition*", they fit within the dimensions of the experience dominated unity of apperception and coping with environmental realities. Krippendorff (2007, p. 8) states: "*To design artifacts for use by others requires second-order understanding*". Science wants to understand the world, art is a creative framework that expresses beauty and values, law is not so much about being right as it is about a human need for justice, medicine is about maintaining the delicate homeostasis of health, while design is about form and Gestalt of a reality translated into an artifact that works for a purpose. According to Krippendorff (2007, p. 3): "*A science defines its subject matter with the aim of developing theories and laws that explains it—excluding the inquiring scientist from these explanations*". Von Foerster on the other

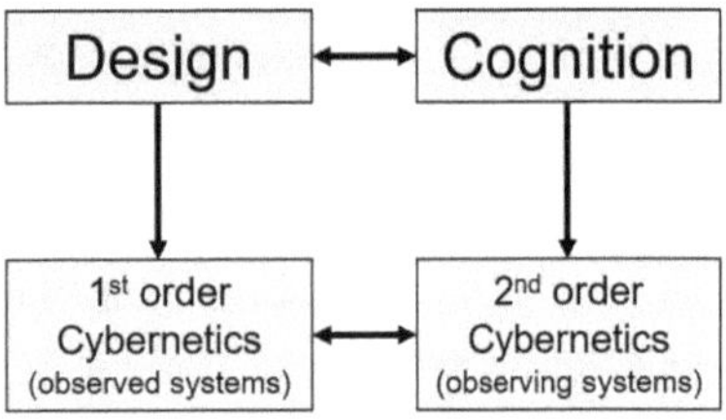

Fig. 15 Design, cognition, and 1st and 2nd order cybernetics. *Source* Own illustration

hand champions the notion that: "*Objectivity is the delusion that observations could be made without an observer*" (Bröcker, 2004, p. 21; Glasersfeld, 1995) (Fig. 15).

Design as stipulated by Simon (1969) was therefore derived to bring this gap towards a harmonic closure by coining it as "*the science of artificial*". Design embraces the bridge between the first-order and second-order stipulation as illustrated above. The foundation of the designer's work is based on the recursive nature of a symmetry between design (first-order) and cognition (second-order) in a hermeneutic value-laden course collecting loop. The notion is highly essential for the designer as an observer, thus it is within the capacity of the designer's ability to reflect as a practitioner that functioning realities emerge (cf. Schön, 1983). Foerster (1979, p. 2): "... *The properties of the observer shall not enter the description of his observations ..., with this I come now to the other root for our cognitive blind spot and this is a peculiar delusion within our Western tradition, namely, 'objectivity*". Foerster (2003, p. 289): "*a brain is required to write a theory of a brain*". Second-order-cybernetics and the constructivist epistemology can contribute much to the design theory as a theory of social-interaction and cognition within the "*design search space*".

6 Team Syntegrity Model

The "*Team Syntegrity Model*" (TSM) also coined as "*Syntegration*" was developed by Beer (1994) and is like the VSM based on the same systemic-logic of bio-inspired cybernetics. The TSM is designed as a model of democratic communication among the different sub-systems (departments) of a larger whole, more specifically for the circulation and integrity of the necessary information for effective team decision making.

The TSM is designed based on self-organization and organizational learning by accelerating cooperation within a viable system (cf. Schwaninger, 2003). In contrast to other similar methods, TSM focuses on the process (cf. Espinosa & Harnden, 2007). Although this requires multiple resources and a strict and tough planning, the empirical experiences illustrate, how it can foster interdisciplinary coordination and collaboration (cf. Schwaninger & Leonard, 2004). The model is designed as a process to create the conditions and the required environment for large group discussion which navigates towards a group decision absolute democratically and without any hierarchies. TSM derives from the terms "*Synergy*", the result of cooperation, and

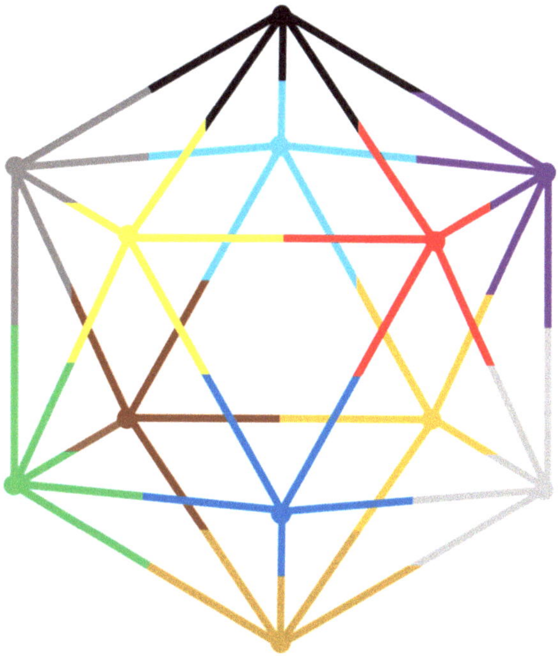

Fig. 16 Icosahedron as organizations embodiment. *Source* An interpretation of Beer (1994)

"*Tensile Integrity*", strength through a fixed structure, which represents the two core aspects of the model. As Beer developed the VSM by looking into the patterns deriving from the human nervous system as the design of an ultra-stability model, the TSM was inspired by the power of platonian geometry via the icosahedron (Fig. 16). The foundation of icosahedron is the holism of its structure, which resembles not a top-down strength of embodiment but moreover a structural strength by design in terms of the aforementioned synergy and tensile integrity of the whole.

The democratic dimension of the TSM starts with the form of the model. Beer chooses the icosahedron because of its special geometrical characteristics. Thus, it is one of five existing regular and convex polyhedrons (platonic solids, Fig. 16), that, according to Beer (1994), perfectly reflect the aspect of democracy: They have no top, no bottom, and no side, and furthermore they are made of regular faces and their edges are congruent. Through this structure, participants move between equally related topics during a Syntegration. Combined with the iterative sessions, the influence of internal group hierarchies is reduced to a minimum during the outcome resolve.

The basis of a solid Syntegration is its unique process, being set in the so called "*protocol*". It is represented through an icosahedron, a polyhedral figure with 20 faces. The 30 edges represent the number of participants called "*infoset*". A Syntegration is always taking place under a certain opening question, which is then divided into 12 (the number of vertices in an icosahedron) different topics by the group. The infoset consists of people invited from the organizer of the Syntegration who are interested to work on the Opening Question. Ideally, they come from different backgrounds and bring in new expertise. For example, in the business environment, these could be representatives from the different functions of a company and other stakeholders. Furthermore, the Syntegration is divided into five stages, as described in a numerical sequence (Table 1).

In the "*Opening*", the process and content of the Syntegration are explained in brief to the infoset by the "*Facilitators*". They are responsible for the fluent process and all organizational aspects. During the "*problem jostle*", the opening question is divided into 12 different topics, done independently by the infoset. Each participant hands in "*Statements of Importance*" (SIs), aspects of the opening question that s/he considers as important. Then, the SIs are clustered and discussed in "*Aggregated Statements of Importance*" by the group. The participant then votes for their preferred topic and the ones with the highest agreement are chosen to be the 12 "*Consolidated Statements of Importance*", the final 12 topics.

The aim of the "*topic auction*" is the optimal composition of the topic groups. Each participant ranks the 12 topics based on individual expertise to see to which topics s/he can contribute the most. An optimization algorithm then assigns the 30 members of the infoset to the 12 topics, each participant to two topics (one edge = two connected vertices of the icosahedron) as the role of an expert, and to two topics as the role of a critic (from the opposite side of the icosahedron).

The "*Outcome Resolve*" is the next step of the process. It is a follow-up of sessions where the teams come together several times to discuss their topics. A team consists of 5 experts and 5 critics, 10 in total. Five is the number of edges that every vertex of

Table 1 Terms of the TSM

Term	Description
Infoset	The total of all participants
Protocol	The pre-planned and strict schedule of the process
Facilitators	The moderators and documentators
1. Opening	Explanation of the protocol to the infoset by the facilitators
2. Problem Jostle	Search and selection of the topics
Statement of importance (SI)	Individual suggestions by the participants for the final topics
Aggregated statement of importance (ASI)	Assorted and structured SIs (e.g. elimination of redundancies)
Consolidated statement of importance (CSI)	The final topics for the outcome resolve, voted by the infoset
3. Topic auction	Arrangement of the infoset to the topic groups
4. Outcome resolve	Group discussions
5. Finalization	Presentation of the group results to the infoset

Source Based on an Interpretation of Beer (1994)

the icosahedron has, a connection of one topic to five other topics. As every participant is an expert of an additional topic and critic in two other topics, s/he gains growing knowledge about the other topics in the other meetings and can bring this information in if a group meets again. Through this iterative process, the knowledge is transferred and integrated automatically into the discussion. In each group meeting, the 5 experts discuss together on the topic while the 5 critics listen. After a certain period, the facilitator interrupts the discussion, and the critics propound their feedback. This happens two or four times during a meeting. The advantage of this mode is that no more than 5 persons are talking at the same time, but still everyone can make contributions. This commits to the efficiency of the communication process, which is normally difficult to achieve in group discussions with more than seven people. Always two meetings take place at the same time, meaning 20 participants are involved as experts or critics. The remaining 10 participants can join the meeting as sole observers but can also have a break. In every meeting, the progress is documented and summarized for the other infoset members.

The "*Finalization*" is the last step of the Syntegration process. Here, the team shares the results of their sessions with the other members. Everybody can now see the gained results and the shared agreement from the other groups. The complete Syntegration process can take up to 3.5 or even 5 days, with meetings between 30 and 90 min.

There are a lot of options for variation during the Syntegration process. Syntegrations have already been based on one of the other four platonic solids (Table 2). This allows a higher flexibility regarding the number of participants and topics, depending on the purpose of a Syntegration and the complexity of the opening question. The same goes for another variable, the timeline. For example, TSM has been shaped to only three hours in total with less iterations and shorter duration (cf. Schwaninger & Leonard, 2004).

Thus, a Syntegration is not limited to a specific application through its structure and protocol contain adjustable variables that allow a wide number of different applications. Beer (1994) himself tested it in the development phase in a business environment, in public and in an educational background. There, TSM has shown to improve the performance of the students (cf. Holmberg, 1997; Martín-Cruz et al., 2014). In addition, it was applied to nonprofit organizations, governments (cf. Leonard, 1996), conferences (cf. Schwaninger & Leonard, 2004) and disaster response management

Table 2 Platonic solids and the TSM

Platonic solid	Participants	Number of topics	Group size
Tetrahedron	6	4	6
Cube	12	8	6
Octahedron	12	6	8
Dodecahedron	30	20	6
Icosahedron	30	12	10

Source Based on an interpretation of Beer (1994)

(cf. Reissberg, 2011). These different settings always require a target group specific design of the protocol but have shown to be successful in its purpose of fast democratic problem-solving, decision making, or knowledge exchange in the context of organizational cognition.

7 The Six Forces Model as a Tool of Designing Complexity

Following the logic of cybernetics, Kamran (2019) extends the Five Forces Model (FFM) (Porter, 1985; cf. Porter, 2008) with an additional holistic force.

Organizations do not operate within a uni-layer milieu, but moreover within a multi-layer of complex web of 9 different layers, which establish the embodiment of design for the "*Six-Forces Model*" (SFM) (Fig. 17). While, the success of the firms is mostly accomplished within the dimensions of "*Economics*", however, the firms must perform very well in a total environment, so that economic rents are achieved over a long period of time by establishing the viability of the firm in advance. Organizations are operationally embedded within a larger "*Total Environment*", by exploitation of the opportunities of the present and exploration of possible worlds of tomorrow, thus designing the notion of organizational ambidexterity. The spheres of the SFM are designed based on the cybernetic logic of recursion, self-organization, and Ashby's Law as illustrated in Table 3. By the visualization of a multi-sphere environment, a design of complexity of operating within the global world of business emerges, which then requires solid variety attenuation models to embrace variety generation by their structural dynamics of absorbing the turbulences of emergence and change. There is a symmetry between control and perturbation of complexity, thus the design of homeostasis requires a design of complexity avant la letter.

The SFM is the most robust strategic model designed based on cybernetic laws, hence viability is the essential question in designing strategy (Table 4). The model is based on:

1. real-time control—as in biological organisms
2. structure is strategy thesis—hence it is within the structural dynamics of variety generating systems, that attenuating variety is absorbed
3. interactional intelligence as the measure of proactive stability—based on the design of embodiment of brain, body, and environment as a whole

Table 3 describes the diverse scientific pillars that define the SFM, whereupon the different streams of application of science contributing to the design of the model are illustrated.

Table 4 delivers a comparative analysis of the SFM in contrast to the diverse mainstream models within the field of management, thus substantiating the diagnostic power of the model, which would be applied to enhance the authors' dimension of designing control via structuring and enhancing the SFM within a geometric holism of the icosahedron as the most stable embodiment of geometric structure. According

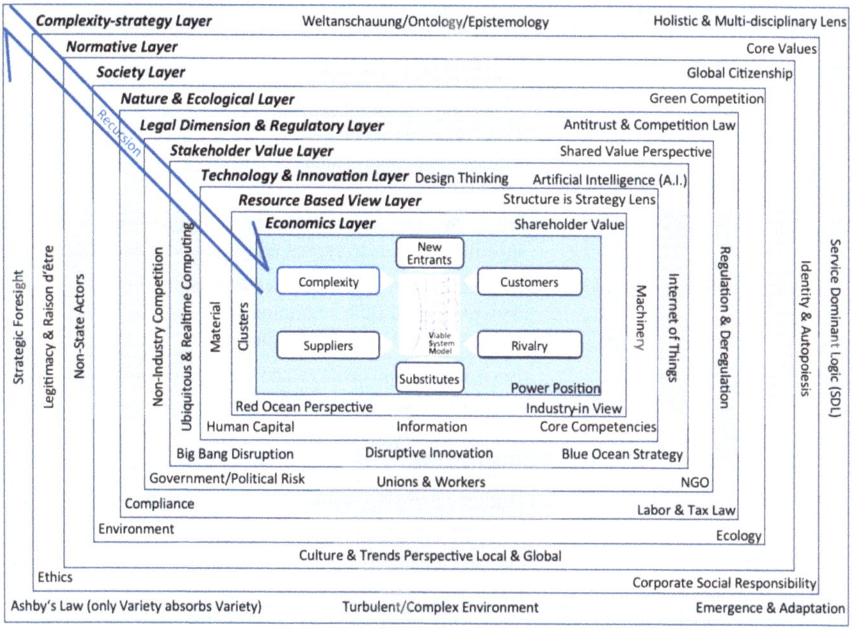

Fig. 17 The SFM embedded in the wider environment from a holistic perspective. *Source* Kamran (2019)

to Beer, who used the term "*reverberation: ideas travel around the icosahedron in all directions, being transformed and becoming progressively less the property of any individual and more that of the infoset as a whole*" (Pickering, 2010, p. 281), thus in a complex setting no force individually captures the essence of turbulence hitting an organization, hence complexity is the embodiment of inter-dependability of forces and interactions in the dynamic of recursion.

8 A Design for Complexity

Figure 18 shows the design process of the new model, following the logic of Ashby's law. As Beer's TSM has already displayed the holism of team collaboration, the icosahedron model has been introduced to structure the complexity within the environment to enable their cognition in an organization. For the question of what layers are adequate to cluster the complexity of the external environment and to mirror them in a firm's internal perspective, the authors decided for an adaptation of the layers of the SFM. As a robust model of capturing complexity, the SFM is a corroborated model, which has paved the way in structuring complexity within a multi-layer design of the environment within the field of strategic management.

Table 3 The layers of the SFM and their contents

Layer	Contents
Economic layer	Shareholder value, industry-in-view, clusters
Normative layer	Core values and ethics, corporate social responsibility, shared value model, identity, autopoiesis, legitimacy and raison d'être
Resource based view layer	RBV, structure is strategy lens, core capabilities, information, human capital and machinery and material
Technology and innovation layer	Blue ocean strategy, disruptive innovation, big bang disruption, real-time communication and ubiquitous computing and internet of things
Nature and ecological layer	Environment and ecology and green competition
Stakeholder value layer	Shared value, unions/workers, non-governmental organizations government and political risk and non-industry
Legal and regulatory layer	Antitrust and competition law, regulation and deregulation, labor and tax law, compliance and intellectual property law
Societal layer	Global citizenship culture and trends perspective, local and glocal and none–state actors
Complex-strategy layer	Weltanschauung/ontology, service dominant logic (SDL), emergence and adaptation, Ashby's law, holistic interdisciplinary lens and strategic foresight

Source Kamran (2019)

Based on Conant Ashby's theorem, the organization must be a model of the milieu where it operates, and therefore creating the conditions for a robust Eigen-value which is capable of formulating adequate strategies to cope with the complexity of the environment. This notion creates the homeostasis of an equation between the complexity of the milieu simulating the complexity within the organization, thus only variety absorbs variety Ashby (1956). Organizational freedom to navigate within a complex environment requires a solid design capability of the organization to create an adequate environmental reality, as modelling is the embodiment of design. Figure 18 describes how a firm designs a reality within the multi-layer design-search-space to navigate an adequate and holistic reality of environmental complexity.

Figure 19 delivers a solid recursive model of how complexity is structured via the icosahedron, while the internal organizational cognition is structured via the VSM. This model delivers a solid structure for reformulating Ashby's Law in terms of: "*only complexity can design complexity.*"

Figure 20 refers to the application of design of complexity in a corporate or educational setting. In a firm, representatives for the different layers come together in the modus similar to Beer's TSM, e.g. to discuss environmental disruption or to decide on strategic changes. The Problem Jostle is still performed, but with a lower freedom of choice as the CSI must fit to the layers. The comparably fast execution of this process through the TSM enables an organization to react contemporarily fast on

Table 4 The SFM in comparison to other strategy models

Model features	Five forces model	Core competencies	Game theory	Competing on the edge	Six forces model
Assumptions	Stable industry structure	Firms as bundle of competencies	Industry view as dynamic oligopoly	Industry in rapid change	Holistic lens total environment
Goal	Defensible position	Sustainable advantage	Temporary advantage	Continuous flow of advantages	Survival immunity viability
Performance drivers	Industry structure	Unique firm competencies	Right moves	Ability to change	Emergent change real-time control
Strategy	Pick and industry and a strategy fit	Create vision build and exploit competencies to realize vision	Make the right competitive and collaborative moves	Gain the "edge" time, pace, shape	Structure is strategy/ dissolve organizational challenges
Success	Profits	Long-term dominance	Short term win	Continual reinvention	Pro-active stability sustainable competitive advantage

Source Kamran (2019)

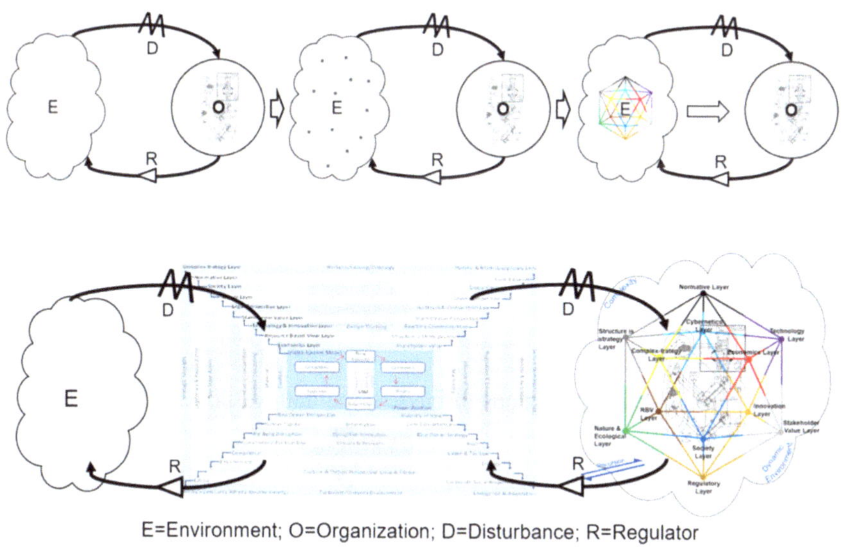

Fig. 18 Development process of the model. *Source* Own illustration

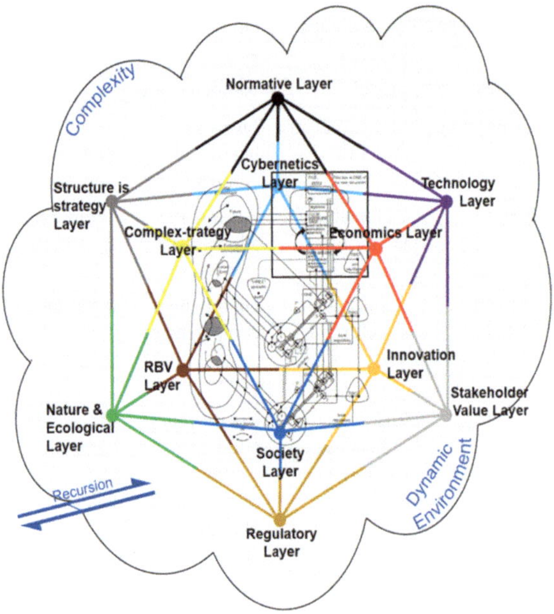

Fig. 19 A design for complexity. *Source* Own illustration

changes within the external environment. As Sect. 6 has shown, the TSM therefore offers options for variation, based on the five platonic solids and depending on the number of participants and the level of complexity of the discussed topic, displayed through the funnel in Fig. 20. For this purpose, the 12 layers can be reduced to make the model adequate.

In addition to the application in the corporate world, the same logic can be applied to engage business students by solving real problem-solving simulations and also in conducting case studies by making sure all essential aspects of the challenge based on a "*design method*" substantiated by the power of cybernetics and group cognition is democratically and holistically pursued and considered.

9 Conclusions

Based on all the broad dimensions discussed, the authors postulate the following propositions:

1. To design is to bringing-forth meaning into a functioning reality of an organization.
2. Functioning is the raison d'être of design, where cybernetics is the science, thus, cybernetics and design reflect each other to deliver the sufficient conditions for effective regulation within contemporary HR management.

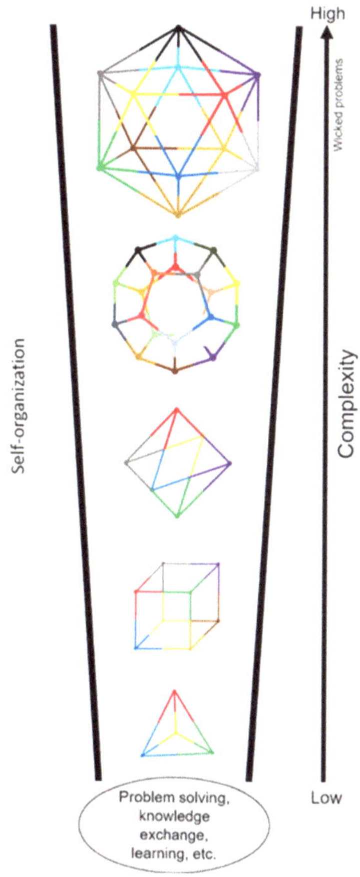

Fig. 20 Application of the model. *Source* Own illustration

3. Hence, all types of organizations require a design of a real system, consisting of a brain, a body, and a cultivating inner-cognitive Eigen-value engaging within an emerging external environment, thus organizational intelligence is embodied, wherein the structural dynamics of an organization is the key to effective teams' management.
4. Design is the collective cognition defined as Eigen-value, of invention, meeting a mutual internally cultivated reality and cohesion of the design team, defined in terms of Eigen-behavior designing an extended market-reality by postulating artifacts, which are resonating back and forth in a co-created meaning between the firm and the customer. Thus, design cannot exist in a vacuum and it exists in a recursive hermeneutical trial and error feedback loop of cybernetics of functionality and phenomenologically bridged via the co-creation of a diverse stakeholder's apperception of value.

5. Artifacts are the outputs of an organizational coherence and Eigen-value was thrown into a co-created reality with the stakeholders. First designers design an organization, which is embedded in an environment that designs artifacts (e.g. customer solutions), whereupon the artifacts deliver semantic solutions to market realities.
6. Design, cybernetics, and cognition are the same phenomenon; hence design's raison d'être is second-order reflection meeting opportunity co-created by a designer in an objective sense of a market resp. environment reality.
7. The nature of design cognition is holistic and requires an organizational design in terms of the "*structure is strategy*" approach, thus the VSM and TSM are delivering the adequate conditions or upon the complexity of modelling and regulation can be effectively absorbed.

The authors have provided a wider outline of diverse cybernetics and design application possibilities. The essential aspects of design as the third pillar of human knowing are extended. The authors have brought sufficient evidence from the sciences of cybernetics (e.g. the VSM, the TSM, and Ashby's Law) to establish a solid bridge between design and cybernetics of first-order (function) and second-order (observer-designer) entering the roam of the "*design search space*". The SFM as a holistic environmental design extended via the icosahedron (TSM) has displayed a solid diagnostic power to structure the complexity of the environment, wherein an organization can navigate by managing complexity more effectively. The discussed "*Good Regulator Theorem*" furthermore substantiates the closure with Archer's dimension of a modelling based circular causality of the first-order and second-order design-embodiment. Furthermore, the research substantiates the laid path by Archer and Simon and delivers many novel approaches to the field, whereupon on a design-based mode of inquiry besides the fields of arts and sciences, many promising adventures of human knowing can be paved.

References

Archer, B. (1968). *The structure of design processes.* Thesis (Ph.D.), The Royal College of Art. https://ethos.bl.uk/OrderDetails.do?uin=uk.bl.ethos.697935

Archer, B. (1976). The three Rs. *Design in General Education,* 8–15.

Archer, B. L. (1979). Design as a discipline. *Design Studies, 1*(1), 17–20.

Ashby, W. R. (1952). *Design for a brain.* Chapman & Hall.

Ashby, W. R. (1956). *Introduction to cybernetics.* Chapman & Hall.

Ashby, W. R. (1958). Requisite variety and its implications for the control of complex systems. *Cybernetica, 1*(2), 83–99.

Ashby, W. R. (1960). *Design for a brain: The origin of adaptive behaviour. Science paperbacks: no. 10.* Springer.

Ashby, W. R. (1962). Principles of the self-organizing system. In H. von Foerster & G. W. Zopf (Eds.), *Principles of self-organization: Transactions of the University of Illinois Symposium* (pp. 255–278). Pergamon Press.

Bayazit, N. (2004). Investigating design: A review of forty years of design research. *Design Issues, 20*(1), 16–29.

Beer, S. (1959). *Cybernetics and management. Management science series*. Wiley.
Beer, S. (1972). *Brain of the firm: A development in management cybernetics*. Lane.
Beer, S. (1979). *The heart of enterprise*. Wiley.
Beer, S. (1981). *Brain of the firm: The managerial cybernetics of organization* (2d ed.). J. Wiley.
Beer, S. (1984). The viable system model: Its provenance, development, methodology and pathology. *Journal of the Operational Research Society, 35*(1), 7–25.
Beer, S. (1985). *Diagnosing the system for organizations. The managerial cybernetics of organization*. Wiley.
Beer, S. (1993). *Designing freedom. The CBC Massey lecture series*. Anansi.
Beer, S. (1994). *Beyond dispute: The invention of team syntegrity. The managerial cybernetics of organization*. Wiley.
Beer, S. (1995). *Diagnosing the system for organizations* (Reprinted [from 1985]). *The managerial cybernetics of organization*. Wiley.
Beer, S. (2000). *Decision and control: The meaning of operational research and management cybernetics* (Reprinted.). *The Stafford Beer classic library*. Wiley.
Beer, S. (2002). What is cybernetics? *Kybernetes, 31*(2), 209–219.
Bernstein, R. J. (2006). The pragmatic century. In S. G. Davaney & W. G. Frisina (Eds.), *The pragmatic century. Conversations with Richard J. Bernstein* (pp. 1–14). State University of New York Press.
Blumrich, M. (1979). Computing nonstandard section properties. *Machine Design, 42*(6).
Bourdieu, P. (1984). *Distinction: A social critique of the judgement of taste*. Harvard Business Review Press.
Bourgine, P., & Stewart, J. (2004). Autopoiesis and cognition. *Artificial Life, 10*(3), 327–345.
Brandom, R. (2013). *Reason, genealogy, and the hermeneutics of magnanimity*. Howison lectures in philosophy. University of California.
Bröcker, M. (2004). The part-of-the-world position of Heinz von Foerster. *Brief Strategic and Systemic Therapy European Review*, (1), 19–26.
Buchanan, R. (1992). Wicked problems in design thinking. *Design Issues, 8*(2), 5.
Christopher, W. F. (2007). *Holistic management: Managing what matters for company success. Wiley series in systems engineering and management: Vol. 46*. Wiley-Interscience.
Conant, R. C., & Ashby, W. R. (1970). Every good regulator of a system must be a model of that system. *International Journal of Systems Science, 1*(2), 89–97.
Cross, N. (1999). Design research: A disciplined conversation. *Design Issues, 15*(2), 5–10.
Cross, N. (2001). Designerly ways of knowing: Design discipline wersus design thinking. *Design Issues, 17*(3), 49–55.
Cross, N. (2006). *Designerly ways of knowing*. Springer.
Cross, N. (2007). *Designerly ways of knowing*. Birkhäuser.
Cross, N. (2011). *Design thinking: Understanding how designers think and work* (Reprinted). Bloomsbury Academic.
Dalsgaard, P. (2014). Pragmatism and design thinking. *International Journal of Design, 8*(1), 143–155.
Davis, K. (2001). *Deconstruction and translation*. St. Jerome.
Derrida, J. (2005). *Paper machine. Cultural memory in the present*. Stanford University Press.
Dewey, J. (1934). *Art as experience*. Perigee books.
Dewey, J. (1938). *Logic: The theory of inquiry*. Henry Holt and Company.
Dreyfus, H. L., & Kelly, S. D. (2011). *All things shining: Reading the Western classics to find meaning in a secular age*. Free Press.
Dubberly, H., & Pangaro, P. (2007). Cybernetics and service-craft: Language for behavior-focused design. *Kybernetes, 36*(9/10), 1301–1317.
Espinosa, A., & Harnden, R. (2007). Team syntegrity and democratic group decision making: Theory and practice. *Journal of the Operational Research Society, 58*(8), 1056–1064.
Fann, K. T. (1970). *Peirce's theory of abduction*. Springer.

Fischer, T. (2014). Wiener's refiguring of a cybernetic design theor. In *2014 IEEE Conference on Norbert Wiener in the 21st Century (21CW)* (pp. 1–7). IEEE.

Foerster, H. (1979). Cybernetics of cybernetics. In K. Krippendorff (Ed.), *Communication and control in society* (pp. 5–8). Gordon and Breach.

Foerster, H. (1981). On cybernetics of cybernetics and social theory. In G. Roth & H. Schwegler (Eds.), *Self-organizing systems* (pp. 102–105). Campus Verlag.

Foerster, H. (1998). *Truth and trust.* Retrieved March 14, 2019, from American Society for Cybernetics: http://www.sympoetic.net/Introduction/truth_and_trust.html

Foerster, H. (2016). Quantum mechanical theory of memory. In C. Pias (Ed.), *Cybernetics—The Macy conferences 1946–1953: The complete transactions* (pp. 98–121). Diaphanes.

Glanville, R. (1994). Variety in design. *Systems Research and Behavioral Science, 11*(3), 95–103.

Glanville, R. (1999). Researching design and designing research. *Design Issues, 15*(2), 80.

Glanville, R. (2004). The purpose of second-order cybernetics. *Kybernetes, 33*(9/10), 1379–1386.

Glanville, R. (2007a). Designing complexity. *Performance Improvement Quarterly, 20*(2), 75–96.

Glanville, R. (2007b). Try again. Fail again. Fail better: The cybernetics in design and the design in cybernetics. *Kybernetes, 36*(9/10), 1173–1206.

Glaserfeld, E. V. (1984). An introduction to radical constructivism. In P. Watzlawick (Ed.), *The invented reality* (pp. 17–40). Norton.

Glasersfeld, E. V. (1995). *Radical constructivism: A way of knowing and learning. Studies in mathematics education series.* Falmer Press.

Glasersfeld, E. V. (1996). Farewell to objectivity. *System Research, 13*(3), 279–286.

Golob, S., & McKay, K. (2016). The Holzwege of Heidegger and Finlay. *Evental Aestetics, 5*(1), 58–87.

Hayek, F. A. V., & Hamowy, R. (Eds.). (2011). *The collected works of F. A. Hayek:/ed. by Bruce Caldwell; Vol. 17. The constitution of liberty: The definitive edition.* Univ. of Chicago Press.

Heidegger, M. (1927). *Sein und Zeit* (19. Auflage (2006)). Max Niemeyer Verlag.

Heidegger, M. (1954). The question concerning technology. *Technology and Values: Essential Readings*, 99–113.

Heidegger, M. (1977). The question concerning technology. *Technologies and Values: Essential Readings*, 99–113.

Holmberg, S. C. (1997). Team syntegrity assessment. *System Practice, 10*(3), 241–254.

Huppatz, D. J. (2015). Revisiting Herbert Simon's "Science of design." *Design Issues, 31*(2), 29–40.

Isaacson, W. (2011). *Steve Jobs: Die autorisierte Biografie des Apple-Gründers* (1st ed.). Bertelsmann.

Isaacson, W. (2015). *The innovators: How a group of inventors, hackers, geniuses, and geeks created the digital revolution.* Simon & Schuster.

James, W. (1898). *Philosophical conceptions and practical results.* University Press.

Johansson-Sköldberg, U., Woodilla, J., & Çetinkaya, M. (2013). Design thinking: Past, present and possible futures. *Creativity and Innovation Management, 22*(2), 121–146.

Kamran, Q. (2013). *Complexity the sixth competitive force that shapes strategy: A cybernetics approach to "Porter's FFM" in turbulent and complex environments.* Lambert Academic Publishing.

Kamran, Q. (2018a). Da-Sein thinking: A phenomenological epistemology for design thinking. *Aesthetics & Philosophy of Art eJournal, 10*(23).

Kamran, Q. (2018b). *The history of design thinking: From pragmatism to phenomenology.* Retrieved November 22, 2018, from SSRN: https://papers.ssrn.com/sol3/papers.cfm?abstract_id=3273872

Kamran, Q. (2019). *Developing a holistic model for competitive strategic management. A critique of Porter's five forces model.* Thesis (Ph.D.), Latvia University.

Kamran, Q., & Neise, R. (2018). *Structure does not follow strategy—Structure is the strategy. How operational excellence through a viable organizational structure delivers the fourth generic strategy: Annual LRN Conference.* Logistic Research Network. Retrieved December 04, 2018, from https://ciltuk.org.uk/LRNfullpapers

Kilpinen, E. (2009). The habitual conception of action and social theory. *Semiotica, 2009*(173), 144.

Krippendorff, K. (1996). A second-order cybernetics of otherness. *Systems Research and Behavioral Science, 13*(3), 311–328.

Krippendorff, K. (2006). *The semantic turn: A new foundation for design.* Taylor & Francis.

Krippendorff, K. (2007). The cybernetics of design and the design of cybernetics. *Kybernetes, 36*(9/10), 1381–1392.

Kruger, J. L. (2004). Translating traces: Deconstruction and the practice of translation. *Literator, 25*(1), 47–72.

Lawson, B. (2006). *How designers think: The design process demystified* (4th ed.). Elsevier/Architectural Press.

Leonard, A. (1996). Team syntegrity: A new methodology for group work. *European Management Journal, 14*(4), 407–413.

Lloyd, P. (2017). From design methods to future-focused thinking: 50 years of design research. *Design Studies, 48,* A1–A8.

Malik, F. (1986). *Strategie des Managements komplexer Systeme: Ein Beitrag zur Management-Kybernetik evyolutionärer Systeme.* Zugl.: St. Gallen, Univ., Habil.-Schr., 1977 (2. Aufl.). *Unternehmung und Unternehmungsführung: Vol. 12.* Haupt.

Malik, F. (2007). *Management: Das A und O des Handwerks* (Aktualisierte und erw. Fassung). *Management:/Fredmund Malik; Bd. 1.* Campus-Verl.

Malik, F. (2015). *Navigieren in Zeiten des Umbruchs: Die Welt neu denken und gestalten* (1st ed.). Campus.

Malik, F., & Probst, G. J. B. (1984). Evolutionary management. In H. Ulrich & G. J. B. Probst (Eds.), *Springer series in synergetics: Vol. 26. Self-organization and management of social systems. Insights, promises, doubts, and questions* (pp. 111–120). Springer.

Martín-Cruz, N., Martín-Pérez, V., Pérez-Ríos, J. M., & Velasco-Jiménez, I. (2014). Team Syntegrity® as a tool for efficient teamwork: An experimental evaluation in a business simulation. *Systems Research and Behavioral Science, 31*(2), 215–226.

Masani, P. R. (1996). Invention: The care and feeding of ideas. *Kybernetes, 25*(9), 70–75.

Maturana, H. R. (1980). Biology of cognition. In H. R. Maturana & F. G. Varela (Eds.), *Autopoesis and cognition: The realization of the living* (pp. 5–58). Springer.

Maturana, H. R., & Varela, F. G. (Eds.). (1980). *Autopoesis and cognition: The realization of the living.* Springer.

McCubbin, N. (2004). Beauty in physics: The legacy of Paul Dirac. *Contemporary Physics, 45*(4), 319–333.

Merlau-Ponty, M. (1974). *The primacy of perception and its philosophical consequences & eye and mind.*

Merriam, C. E. (1925). *New aspects of politics.* Univ. of Chicago Press.

Mitchell, M. (2009). *Complexity: A guided tour.* University Press.

Ostman, L. (2005). Design theory is a philosophical discipline—Reframing the epistemological issues in design theory. In *EAD 06 Proceedings of the 6th International Conference of the European Academy of Design* (pp. 1–13).

Owen, C. (2007). Design thinking: Notes on ins nature and use. *Design Research Quarterly, 2*(1), 16–27.

Paczuski, M., & Bak, P. (1999). *Self-organization of complex systems.* https://arxiv.org/pdf/cond-mat/9906077

Pangaro, P. (2016). Why do we want to live in cybernetics? *Cybernets and Human Knowing, 23*(2), 9–21.

Pask, G. (1961). *A proposed evolutionary model.* Reprinted in H. von Foerster & G. Zopf (Eds.), Principles of self-organisation. Pergamon Press.

Peirce, C. S. (1877). The fixation of belief. *Popular Science Monthly, 12*(November), 1–15.

Peirce, C. S. (1878). How to make our ideas clear. *Popular Science Monthly, 12*(January), 286–302.

Pfeifer, R., & Bongard, J. (2007). *How the body shapes the way we think: A new view of intelligence.* MIT Press.
Pfeifer, R., & Scheier, C. (2001). *Understanding intelligence.* MIT Press.
Pickering, A. (2010). *The cybernetic brain: Sketches of another future.* Univ. of Chicago Press.
Popper, K. (2012). *In search of a better world: Lectures and essays from thirty years.* Taylor and Francis.
Porter, M. E. (1985). *Competitive advantage: Creating and sustaining superior performance.* Free Press.
Porter, M. E. (2008). The five competitive forces that shape strategy. *Havard Business Review, 86*(1), 78–93.
Reiners, G. M. (2012). Understanding the differences between Husserl's (descriptive) and Heidegger's (interpretive) phenomenological research. *Journal of Nursing & Care, 01*(05).
Reissberg, A. (2011). The advanced syntegration as the most effective and efficient tool for large-scale disaster response coordination. *Systems Research and Behavioral Science, 28*(5), 455–464.
Rittel, H. W. J., & Webber, M. M. (1973). Dilemmas in a general theory of planning. *Policy Sciences, 4*(2), 155–169.
Rocha, L. M. (1996). Eigenbehavior and symbols. *Systems Research, 13*(3), 371–384.
Romme, A. G. L. (2003). Making a difference: Organization as design. *Organization Science, 14*(5), 558–573.
Rorty, R. (1980). *Philosophy and the mirror of nature* (1. paperback print). Princeton University Press.
Rorty, R. (1999). *Philosophy and social hope. A Penguin original.* Penguin Books.
Rylander, J. (2012). Using levels of inquiry in the classroom. In K. F. Nettleton & L. C. Lennex (Eds.), *Advances in educational technologies and instructional design. Cases on inquiry through instructional technology in math and science* (pp. 1–20). IGI Global.
Schön, D. A. (1983). *The reflective practitioner: How professionals think in action.* Temple Smith.
Schwaninger, M. (2003). A cybernetic model to enhance organizational intelligence. *Systems Analysis Modelling Simulation, 43*(1), 53–65.
Schwaninger, M. (2006). Design for viable organizations: The diagnostic power of the viable system model. *Kybernetes, 35*(7/8), 955–966.
Schwaninger, M. (2010). Model-based management (MBM): A vital prerequisite for organizational viability. *Kybernetes, 39*(9/10), 1419–1428.
Schwaninger, M., & Leonard, A. (2004). A dialogue on the future of ISSS: Team syntegrity sessions at the crete conference. *Systems Research and Behavioral Science, 21*(5), 529–537.
Sheehan, T. (2014). What, after all, was Heidegger about? *Continental Philosophy Review, 47*(3–4), 249–274.
Simon, H. A. (1962). The architecture of complexity. *Proceedings of the American Philosophical Society, 106*(6), 467–482.
Simon, H. A. (1969). *The sciences of the artificial.* The MIT Press.
Swingewood, A. (1984). *A short history of sociological thought.* Macmillan.
Thomas, J. C., & Carroll, J. M. (1979). The psychological study of design. *Design Studies, 1*(1), 5–11.
Ulrich, H., & Probst, G. J. B. (Eds.). (1984). *Springer series in synergetics: Vol. 26. Self-organization and management of social systems: Insights, promises, doubts, and questions.* Springer.
Varela, F. (1984). Two principles for self-organization. In: H. Ulrich & G. J. B. Probst (Eds.), *Self-organization and management of social systems.* Springer Series
von Foerster, H. (1992). *Einführung in den Konstruktivismus.* Piper.
von Foerster, H. (2003). *Understanding.* Springer.
Watzlawick, P. (1984). *The invented reality: How do we know what we believe we know? Contributions to constructivism.*
Wiener, N. (1948). *Cybernetics; or control and communication in the animal and the machine.* John Wiley.

Promoting Gender Equality in Companies: A Path Forward in Building a More Responsible Future—Diagnosis of the "*Happy Company Group*"

Luís Miguel Martins Sá and Carolina Feliciana Machado

Abstract Social Responsibility, in particular, the Promotion of Gender Equality, has been an element with increasing relevance with regard to the prestige and reputation of organizations (Fundação Instituto de Administração, 2019), not only today, but also, and increasingly, in the future. It is considered a fundamental right and an essential component for economic growth (Fundação Instituto de Administração, 2019). The implementation of a formalized gender equality policy must be based on positive factors such as the capitalization of resources, loyalty and motivation, productivity, and reputation and image, both internally and externally (Comissão para a Cidadania e Igualdade de Género, 2008). Likewise, the evaluation of a company's level of promotion of gender equality involves analyzing nine dimensions of Gender Equality in the internal and external environment of the organization, namely, the company's mission and values; recruitment and selection of personnel; lifelong learning; remuneration and career management; social dialogue and participation of workers; respect for human dignity; information, communication and image; conciliation between professional, family and personal life; and maternity and paternity protection and family assistance (Comissão para a Cidadania e Igualdade de Género, 2008).

In this way, the present work consists of an assessment of the *Happy Company Group*'s level of promotion of gender equality, through the application of the diagnostic questionnaire produced within the scope of the "Social Dialogue and Equality in Companies" project financed by the European initiative EQUAL (Comissão para a Igualdade no Trabalho e no Emprego et al., 2008). The questionnaire

This is a fictitious name to guarantee anonymity.

L. M. M. Sá · C. F. Machado (✉)
School of Economics and Management, University of Minho, Braga, Portugal
e-mail: carolina@eeg.uminho.pt

C. F. Machado
Interdisciplinary Centre of Social Sciences (CICS.NOVA.UMinho), University of Minho, Braga, Portugal

C. Machado (ed.), *Building the Future with Human Resource Management*,
Management and Industrial Engineering, https://doi.org/10.1007/978-3-031-52811-8_9

was completed based on document analysis and through informal interviews with company employees.

Keywords Gender equality · Promotion of gender equality · Social responsibility · Socially responsible HRM · European context · Portuguese context

1 Introduction

Social Responsibility is an element with increasing relevance with regard to the prestige and reputation of organizations (Fundação Instituto de Administração, 2019). One of the reasons for this growing relevance, as the Fundação Instituto de Administração (FIA) (2019) suggests, is the market itself. Consumers no longer guide their decisions solely based on quality or price, but rather on the seriousness and commitment of brands with society (Fundação Instituto de Administração, 2019). However, corporate social responsibility is a two-dimensional concept and, as Froes and Melo Neto (2004) list, it is necessary to approach the concept from two dimensions: the internal public and the community. In this sense, the SR of an organization must be created, initially, through the involvement of all its members, in order to legitimize the participation of the remaining stakeholders (Turker, 2018).

Likewise, Gender Equality has been an urgent and much debated issue. As suggested by the Constituição da República Portuguesa (1976), specifically in article 13 the gender equality can be seen as the "equality between men and women with regard to their rights, treatment, responsibilities, opportunities and economic and social achievements" and can only be achieved when men and women enjoy the same rights and the same responsibilities and opportunities in all sectors of society and when the different interests, needs and priorities of men and women are valued equally (Kroon & Paauwe, 2014).

The promotion of gender equality in a company is based on nine dimensions in the internal and external environment of the organization, namely, the company's mission and values; recruitment and selection of personnel; lifelong learning; remuneration and career management; social dialogue and participation of workers; respect for human dignity; information, communication and image; conciliation between professional, family and personal life; and protection in maternity and paternity and family assistance (Comissão para a Cidadania e Igualdade de Género, 2008).

2 Literature Review

2.1 *Social Responsibility and Corporate Social Responsibility*

Social Responsibility is an element with increasing relevance with regard to the prestige and reputation of organizations (Fundação Instituto de Administração, 2019). One of the reasons for this growing relevance, as suggested by FIA (2019), is the market itself. Consumers no longer guide their decisions solely based on quality or price, but rather on the seriousness and commitment of brands to society. Another argument suggested by the FIA is part of the Stakeholder Theory proposed by Freeman (2010). Under this perspective, organizations are inserted in an environment made up of several stakeholders—suppliers, customers, community, partners, environment, governments and other stakeholders—who are directly or indirectly affected by their actions (Freeman, 2010). Costs and benefits arise from this interaction. Therefore, the condition of these stakeholders also, and necessarily, affects the condition of the company itself (Freeman, 2010). Consequently, the better the condition of these stakeholders and the better the company is able to benefit them, the better the outcomes will be for itself (Fundação Instituto de Administração, 2019).

In the words of Neto (2010), the existence of organizations requires the approval of the organized civil society. Organizations are expected to contribute socially and environmentally in a positive way, to legitimize their actions. This vision essentially focuses on the actions carried out by organizations in relation to their external environment (Neto, 2010). Likewise, de Almeida (2022) sees corporate social responsibility (CSR) as an organization's commitment to society, in the form of acts and attitudes that positively affect it, or some group within that society 5. However, this one-dimensional view assumes that CSR only relates to the external dimension of the organization and marginalizes its internal dimension (Froes & de Melo Neto, 2004).

In contrast, Froes and de Melo Neto (2004) highlight the need to approach the concept in its two dimensions: the internal public and the community. According to this perspective, it is necessary to look at the way in which organizations promote social responsibility internally, among their members. There is a relationship between these two dimensions. As Turker (2018) highlights, the creation of work and the training of employees are considered the most positive impacts of companies on society. It would be mere hypocrisy for organizations to carry out actions aimed at the sustainable socioeconomic development of society and not do the even with their collaborators.

In this sense, social responsibility of an organization must be created, initially, through the involvement of all its members, in order to legitimize the participation of the remaining stakeholders (society) (Neto, 2010). Likewise, according to the previously named Commission of the European Communities, corporate social responsibility can be understood as "a concept whereby companies integrate social and environmental concerns in their business operations and in their interaction with their stakeholders on a voluntary basis" (Comissão Europeia, 2001: 66). Based on this vision, it will be necessary to point out that, in the eyes of the public externally, the organization's employees will always be the best witness to its values and ethical behaviors (Neto, 2010). Consequently, employees must be at the top of the CSR agenda in any organization (Turker, 2018). In the same vein, incorporating the two dimensions of CSR, in 1999, the World Business Council for Sustainable Development presented, in 1999, a definition of CSR: "The commitment of business to contribute to sustainable economic development working with employees, their families, the local community and society at large to improve their quality of life" (World Business Council for Sustainable Development, 1999, referred by Dahlsrud, 2008: 7).

However, as Dobers (2009) warns, CSR does not only address relationships between companies and other actors, but also as a normative content that addresses the responsibilities that companies may have in our changing social and economic context. According to Comissão Europeia (2011), this regulatory framework is often guided by the main international guidelines, such as the Organization for Economic Co-operation and Development (OECD) Guidelines for Multinational Enterprises, the ten principles of the UN Global Pact, the International Organization for Standardization (ISO) with the Guidance Standard on Social Responsibility, the International Labor Organization, the Tripartite Declaration of Principles Relating to Multinational Enterprises and Social Policy, the UN Guiding Principles on Business and Human Rights, among others.

In this way, considering the two-dimensionality and guiding nature of the concept, this work will consider CSR as the normative framework that guides the actions of a company, with a view to promoting the sustainable socioeconomic development of its members and the society in which it operates (de Almeida, 2022; Fundação Instituto de Administração, 2019; Neto, 2010; Turker, 2018). This definition of CSR implies that a socially responsible company is one that rethinks its current positions, behaviors and conduct and, in this way, articulates itself with the objective of putting into practice attitudes that promote the well-being of those involved, society and of the environment (Fundação Instituto de Administração, 2019).

2.2 Socially Responsible Human Resources Management Practices

Once the concept has been defined and given the objective of the present study, we will seek to understand how social responsibility and its inherent concerns materialize, or not, in the 'modus operandi' of companies, specifically, through its integration into human resource management practices.

In its broadest sense, social responsibility materializes in companies through compliance with ISO standards in matters of ethics and social responsibility, in the implementation of codes of ethics and conduct, in compliance with laws and standards in matters of data protection and in the integration of ethics in the company's strategic development and in its relationships with different stakeholders (Universidade Europeia, 2023). However, the best way to understand how companies promote social responsibility among their employees is to look at their human resources management (HRM) practices (de Almeida, 2022).

For companies, there is no formal or single way to integrate social responsibility into HRM (Turker, 2018). This process can, as Turker indicates, be driven by the obligations imposed by the legal framework. Furthermore, civil society organizations can also have an influence on this process. These have gained increasing influence on HRM activities, as they raise society's awareness of problematic situations in companies (Turker, 2018).

The concept of Human Resources (HR) can be understood as the workers who operate in a given organization and who, in this way, contribute to its functioning, development and decision-making process (Michaels et al., 2001). Its management is carried out through various practices that may have different applications, actors and repercussions (Cardoso, 2012). In this way, HRM covers practices such as recruitment and selection, performance evaluation, remuneration, benefits, recognition/awards, training, development plans, talent management, among others (Cardoso, 2012).

"Organizations with more developed HRM practices also have better CSR policies..." (Turker, 2018). The Comissão Europeia (2011) also relates the issue of CSR to practices at work, namely with regard to training, diversity, gender equality and health and well-being of employees.

However, it is necessary to clarify what we understand by socially responsible human resources management practices (SRHRMP). On the one hand, from a legal point of view, socially responsible human resource management (SRHRM) relates to freedom of association and the effective recognition of the right to demand better wages and working conditions through professional organizations, elimination of all forced forms of work, abolition of child labor and elimination of discrimination related to employment or occupation (International Labour Organisation, 2009). On the other hand, as Turker (2018) indicates, several academics have tried to establish the basic principles of a SRHRM. Among them, Kroon and Paawe (2014) suggested three areas of social responsibility in labor practices, namely, the right to freedom in terms of contract duration and the nature of the employment relationship; the right to well-being in terms of access to the exercise of labor rights, career protection and development and a socially and physically healthy working environment; and the right to equality in terms of labor policies and benefits. It is mainly on this last point that this chapter intends to focus.

Enshrined in Article 13 of the Constitution of the Portuguese Republic, the "Principle of Equality" establishes that "All citizens have the same social dignity and are equal before the law" and that "no one can be privileged, benefited, harmed, deprived of any right or exempt of any duty due to ancestry, sex, race, language, territory of origin, religion, political or ideological convictions, education, economic situation, social condition or sexual orientation" (Lei Constitucional, art. 13°, 2005). Likewise, with regard to companies, Código de Trabalho (Lei n° 7/2009, 2009), as expected, also addresses the issue of Equality and Non-Discrimination. In this matter, it establishes equality in access to employment and work; the prohibition of discrimination;

the prohibition of harassment; equality and non-discrimination based on sex, namely, in access to employment, professional activity and training; equal working conditions, in particular, remuneration; and the obligation to keep a record of recruitment processes, for 5 years, with disaggregation by sex (Lei nº 7/2009, 2009).

However, gender equality is a complex topic and limiting it to the legal framework constitutes a naive and unrealistic approach.

2.3 *Gender Equality*

Gender Equality has been an urgent and much debated issue in the field of public policies, mainly in Europe. This can be seen as equality between men and women in terms of their rights, treatment, responsibilities, opportunities and economic and social achievements (Diretiva 2006/53/CE, 2006) and can only be achieved when men and women enjoy the same rights and the same responsibilities and opportunities in all sectors of society and when different interests, needs and priorities of men and women are valued equally (Eurostat, 2021).

2.3.1 The European Context

In the European context, gender equality is considered a fundamental right, an essential component for economic growth and a key principle of the European Pillar of Social Rights (Eurofound, 2023). However, despite European efforts on gender equality and the progress achieved, namely, legislation for equal treatment, gender integration, integration of the gender perspective in all other policies and specific measures with regard to employment, wages, working conditions and quality of life, there is still a lot to do (Eurofound, 2023).

Regarding employment, according to Eurostat (2019), in 2017, the employability rate of women between 20 and 64 years old (66.5%) was 11.5 p.p. lower than the employability rate of men of the same age (78%), in the EU. Furthermore, in 2020, it was observed that women's gross hourly earnings were, on average, around 13.0% lower than men's in the EU (Eurostat, 2021).

It was in this context that the European Commission launched, in 2020, a Gender Equality Strategy 2020-2025, which addresses issues such as violence against women, salary transparency and the gender pay gap, the balance between men and women on company boards and reconciling professional life and family life. As a result of this strategy, the Commission proposed binding measures on pay transparency the following year. Furthermore, the European Parliament held, in October 2021, the second European Week for Gender Equality, which was the stage for several debates on issues related to gender equality and where the European Institute for Gender Equality (EIGE) published the findings of its 2021 Gender Equality Index (Eurostat, 2021).

2.3.2 The Portuguese Context

The principle of gender equality is also clearly present in the Portuguese legislation. In addition to article 13º of the Constituição da República Portuguesa (2005) mentioned previously, article 9º also lists the importance of the State in promoting equality between men and women. Other articles of the same document, namely, article 26º, which reads "The rights to personal identity, personality development, civil capacity, citizenship, good name and reputation, image, word, to the reserved intimacy of private and family life and to legal protection against any forms of discrimination.", article 47º "Everyone has the right to freely choose the profession or type of work, subject to legal restrictions imposed by the collective interest or inherent to their own capacity." and article 58º "Equal opportunities in the choice of profession or type of work and conditions so that access to any position, work or professional category is not prohibited or limited, depending on sex.", emphasize the concern in guaranteeing of gender equality (Constituição da República Portuguesa, 2005).

The Labor Code (Lei nº 7/2009, 2009) also highlights issues of gender equality in its Subsection III, through articles 25º, 30º and 31º.

Furthermore, within the scope of the III National Plan for Equality, Citizenship and Gender, which includes a set of actions specifically aimed at promoting equal treatment and opportunities between men and women in the labor market, a Guide for the Implementation of Equality Plans in Companies was published (Comissão para a Cidadania e Igualdade de Género, 2008). This guide serves as a tool for the implementation, mainstreaming and operationalization of gender equality in a business context (Comissão para a Cidadania e Igualdade de Género, 2008).

In 2018, the National Strategy for Equality and Non-Discrimination 2018-2030 Portugal + Igual was also approved. The action plan of this strategy provides for measures aimed at promoting equal pay, reconciling professional, family and personal life, preventing and combating violence against women and domestic violence, combating discrimination related to sexual orientation, gender identity and sexual characteristics and inclusion of innovative areas such as scientific and technological development and the mainstreaming of the intersectional perspective and multiple discrimination (Santos, 2022).

As a result of these efforts, "gender equality in Portugal has greatly evolved positively in all aspects" (Brands' Trabalho, 2020: 1), namely through the increase in parental leave, the reduction in women's illiteracy rates, the reduction in the wage discrepancy between men and women in the same role (Brands' Trabalho, 2020).

2.3.3 Gender Equality in Companies

In line with the European and national legislation and guidelines, companies must also guarantee the same opportunities and conditions for women and men, in particular, at a professional level. Not only because it is a legal imperative, but because gender equality is characterized as a central element for sustainable development.

According to the Comissão para a Igualdade no Trabalho e no Emprego et al. (2008), the implementation of a formalized gender equality policy can be based on the following positive factors:

Internal Level:

Capitalization of Resources—(a) Enhancing the skills and knowledge of its human resources, by including everyone.

Loyalty and Motivation—(a) Loyalty of workers; (b) Contribution to a better working environment; (c) Increased motivation.

Productivity—(a) Reduction in absenteeism; (b) Reduction in turnover rates; (c) Greater satisfaction of workers.

Reputation and Image—(a) Greater transparency in relationships with workers and their organizations;

External Level:

Capitalization of Resources—(a) Innovation and creativity in the products and solutions made available to customers and the market.

Loyalty and Motivation—(a) Attraction and retention of human capital; (b) Trust and credibility in the market and the community in which it operates.

Productivity—(a) Competitiveness factor due to greater profitability of resources.

Reputation and Image—(a) Recognition for promoting gender equality measures, as a company with good practices within the framework of social responsibility; (b) Increased brand awareness and value.

3 Methodology

The present study starts from a qualitative approach, by carrying out a diagnosis of the situation of the *Happy Company* in the field of equality between men and women and, inherently, the reconciliation of professional, family and personal life and protection of motherhood and fatherhood.

As such, the present study was carried out based on the questionnaire produced within the scope of the "Social Dialogue and Equality in Companies" project financed by the European initiative EQUAL (Comissão para a Igualdade no Trabalho e no Emprego et al., 2008). The questionnaire (implemented as informal interviews to all employees) consists of 65 questions with an alternative and exclusive answer between "Yes" or "No", covering 9 dimensions of Gender Equality in the organization's internal and external environment, namely:

- Company mission and values;
- Recruitment and selection of personnel;
- Lifelong learning;
- Remuneration and career management;
- Social dialogue and participation of workers and/or their representative organizations;
- Duty of respect for the dignity of women and men in the workplace;

- Information, communication and image;
- Reconciling professional, family and personal life;
- Maternity and paternity protection and family assistance.

Analyzing the results of the questionnaire made it possible to diagnose the company's position in terms of gender equality.

To answer the questionnaire, the participated observation methodology was used, complemented with document analysis. Participated observation, according to Eriksson and Kovalainen (2008), is one of the most demanding forms of collecting empirical materials and requires the researcher to participate in the culture or context that is being observed. However, due to the limitations of this method, a documentary analysis was also carried out, having collected several documents, including the company's organizational chart, management reports, action plans, code of ethics and conduct.

The mixture of the two methods made it possible to identify not only the standards and guidelines that the company defined as part of its gender equality policy, as well as their effective application.

4 The *Happy Company* Group

4.1 Characterization

Happy Company is a business group made up of 6 companies, the parent company *Happy Company* Group and 5 subsidiaries, namely Financial/ Management Support City A, Financial/ Management Support City B, Strategic, Human Capital and BlueSky. The group provides consultancy services in the area of management and business, real estate mediation and leasing of real estate assets. Currently, the company has 33 employees aged between 23 and 45 years old.

Mission

The company assumes that its mission is to transform knowledge into value, promoting organizational development and improving competitiveness and business productivity.

Vision

The company aims to be a reference in the provision of management services, by creating added value for companies, through a strong commitment in the area of R&D+I and the continuous training of human capital, towards Excellence.

Values

The company assumes that its values are: knowledge and innovation; commitment and competence; team spirit and dynamism; trust and honesty; and, ethical, social and environmental responsibility.

4.2 Diagnosis

4.2.1 Company Mission and Values

One of the values based on the *Happy Company Group's* policy is ethical, social and environmental responsibility. In this sense, the company looks at equality between men and women as a priority issue for its own organizational development and as a catalyst for the creation of value.

It is important to note that the company includes the principle of gender equality in its strategic documents, in particular, in the Code of Ethics and Conduct, and in the Internal Regulations. Furthermore, the company has participated in several projects that pursue this same objective, contributing to the awareness and dissemination of these values, in particular, the "Social Dialogue and Equality in Companies" project.

More recently, in May 2022, *Happy Company* joined the Forum of Organizations for Equality (iGen), this forum is characterized as a group of organizations, national and multinational, that have made a commitment to reinforce and highlight their organizational culture of social responsibility, incorporating into its strategies and management models the principles of equality between men and women at work and employment.

The company has also established an open participation policy, in which all its employees, regardless of gender, or other characteristic that distinguishes them, are invited to participate in decision-making processes, contributing critically to the company's direction. This policy allows all contributions from its members to be absorbed and, in the form of debate, allows decisions and solutions to be reached that integrate different visions and, for this reason, are considered more mature.

Finally, it is important to highlight that the company has an action plan for gender equality in place, which aims to avoid and mitigate any discrimination based on gender.

4.2.2 Recruitment and Selection of Personnel

Happy Company deeply believes that the value of people is not evident through their characteristics, but rather in their actions and the attitude they take at work and towards others. Likewise, when it is necessary to recruit a male or female employee to reinforce the team, gender is not a factor to consider.

The recruitment team, during the recruitment and selection process, focuses on the technical and transversal skills of the candidates and seeks to assess their alignment with the organization's philosophy.

The constitution of the *Happy Company* team does not follow the principle of gender parity for a very simple reason. Believing in individual merit and skills, in all recruitment processes, what prevails are skills and alignment with the company's philosophy, therefore not considering the issue of gender. Furthermore, these processes do not contain any indirectly discriminatory element based on gender and comply with current legislation, namely the storage of information, by gender, relating to the recruitment and selection processes carried out in the last five years.

4.2.3 Lifelong Learning

Happy Company team's Training Plans are formulated taking into account the company's technical and operational needs and the career progression expectations of its employees. As such, plans are made individually, in consensus with the employee in question, without discriminating against anyone with regard to qualification opportunities, career progression and access to management positions.

The company has a training module that certifies equality between women and men, training that is also taught to other entities by the *Happy Company* team. The training highlights the importance of gender equality in organizations, the need to integrate this topic into Codes of Ethics and Conduct, and an action plan designed to deal with issues of workplace harassment.

Furthermore, the company offers weekly or fortnightly training sessions in which all company employees are invited to participate and, often, invited to provide the training themselves.

Furthermore, the company guarantees all employees equal access to the minimum number of hours of certified training established by law.

4.2.4 Remuneration and Career Management

The *Happy Company Group* ensures compliance with the principle of equal pay for equal work or work of equal value, in its remuneration policy. As with recruitment processes, merit and skills of the worker are considered, with gender not being a relevant factor.

The same applies to the award of additional remuneration. As a general rule, work is distributed equitably and, whenever there is an increase in the volume of work, the awarding of bonuses and incentives follows the principle of the volume of work supported by each worker, and is therefore attributed complementary remuneration based on the work performed by a worker.

When *Happy Company* appoints a worker to a management position, it takes their skills into account, selecting the person who meets the best conditions to take on the position. It is important to note that although the Members of the Statutory Bodies are mostly men, the leadership positions are, for the most part, headed by women. Furthermore, the company has a performance assessment system based on objective criteria, in which the level of performance is proportionally related to the degree of achievement of these objectives. The objectives are common between men and women, considering individual specificities, time spent at work and the ambition of the employee. As such, there is no discrimination based on gender.

4.2.5 Social Dialogue and Participation of Workers and/or Their Representative Organizations

As previously mentioned, *Happy Company* has an open participation policy, in which all its workers are invited to make suggestions and look at the company and its way of operating critically. The same happens in relation to gender equality measures, reconciliation between professional, family and personal life and the protection of maternity and paternity. In order to integrate the suggestions of its members, the company promotes meetings in which these issues are addressed in order to integrate these concerns into the organizational culture and promote the happiness and well-being of its members.

Furthermore, these questions are of significant importance when the annual satisfaction questionnaires of the group's male and female workers are carried out.

4.2.6 Duty of Respect for the Dignity of Women and Men in the Workplace

Happy Company has a Code of Ethics and Conduct that incorporates a set of measures and standards that aim to ensure respect for the dignity of women and men in the workplace.

All of these issues are addressed with a high degree of seriousness within the organization, with the document providing for a formal procedure for filing complaints in the event of discrimination based on sex and/or workplace harassment.

After filing a complaint, all the facts are investigated, safeguarding the elements involved until a definitive conclusion is reached.

Depending on the seriousness of the case and the facts, proportional measures are adopted aimed, firstly, at safeguarding the victim and, subsequently, at adopting punitive measures for the "aggressor".

4.2.7 Information, Communication and Image

In all internal and external communication, *Happy Company* uses inclusive and neutral grammatical forms, in accordance with the principle of equality and non-discrimination based on gender.

In advertising measures, in particular, in the dissemination of graphic content, there is special attention to gender parity, always seeking to ensure an equal representation of men and women in publications.

Furthermore, the company gives to all its workers, suppliers, and customers, information regarding the rights and duties of male and female workers, in matters of equality and non-discrimination based on gender, maternity and paternity.

4.2.8 Reconciling Professional, Family and Personal Life

Happy Company has a wide portfolio of clients which, seasonally, leads to a significant increase in work. Therefore, sometimes the excessive increase in work makes it difficult to reconcile professional, family and personal life, as it demands extra effort from male and female workers. However, the company has an Action Plan in place to deal with this situation, as a result of the recognition of the extreme importance that this factor has for its workers. In this sense, the company provides some flexibility in working hours, allowing workers to adapt their working hours, depending on their needs, as long as the company's operational objectives are not put into question. Furthermore, the company offers its male and female workers a mixed work regime, allowing a third of working hours to be done remotely (teleworking). In order to avoid additional expenses for its members, the company provides equipment, namely monitors, orthopedic chairs, footrests, computer and office material, so that working conditions at home are exactly the same as those offered at their Open Office.

The company also has a wide network of partners, from the health, aesthetics, catering, clothing sales, among others, with which it has established protocols that provide discounts for its employees when using services and purchasing products.

Additionally, the company promotes activities linked to health promotion, including the importance of healthy eating and physical exercise.

4.2.9 Maternity and Paternity Protection and Family Assistance

Happy Company workers are aged between 23 and 45, and most are at an age that is considered biologically "appropriate" to have children. Furthermore, many of the members have had children since joining the team.

The company views motherhood and fatherhood naturally, always aware that it is a very important stage for the personal fulfillment of its workers. As such, *Happy Company* offers full support to its members during this phase, inaugurating it with a celebration: the Baby Shower. In this activity, the team meets and purchases a series of utilities that will be essential during the first months of this stage, and, in the form of a get-together, a small celebration of the event is held, where the goods are handed over to the future father or mother.

Furthermore, the company encourages men to take the 15 days of parental leave provided for by law, which are remunerated by Social Security and, if there is a need to reinforce the team, hires to replace male or female workers on parental leave.

5 Conclusion

It is undoubted that equality and non-discrimination based on gender are extremely important factors for organizations, not only at an external level and in the image that companies convey to society, but also for the justice, happiness and well-being of their employees internally.

After carrying out a gender equality diagnosis at the *Happy Company Group*, using documentary analysis and informal interviews, it was possible to conclude that gender equality is a basic principle of the culture that makes up the organization. This is formally reflected in internal documents, as well as in day-to-day practices. The integration of *Happy Company* into the iGen Forum of Organizations for Equality was a recognition of this fact.

Furthermore, it was also possible to identify the administration's concern and awareness of the 9 dimensions of gender equality in the organization's internal and external environment.

In short, we can conclude that the *Happy Company Group* not only asserts itself as an ethical and socially responsible organization, but also implements social responsibility practices at all levels of the organization.

References

Brands' Trabalho. (2020). Igualdade de género nas empresas. Retrieved 30 October, 2023, from Https://Eco.Sapo.Pt/2020/03/09/Igualdade-de-Genero-Nas-Empresas/.

Cardoso, L. (2012). *Práticas de gestão de recursos humanos e desempenho organizacional* [Dissertação Doutoral]. Instituto Superior de Psicologia Aplicada (Portugal).

Comissão Europeia. (2001). *GREEN PAPER Promoting a European Framework for Corporate Social Responsibilities*. Retrieved 30 October, 2023, from https://ec.europa.eu/commission/presscorner/detail/en/DOC_01_9.

Comissão Europeia. (2011). *A renewed EU strategy 2011–14 for Corporate Social Responsibility*. Retrieved 30 October, 2023, from https://www.eumonitor.eu/9353000/1/j4nvkkpftveemt7_j9vvik7m1c3gyxp/vitwrzhm31gk.

Comissão para a Cidadania e Igualdade de Género. (2008). *Guião para a Implementação de Planos de Igualdade nas Empresas*. Retrieved 4 November, 2023, from chrome-extension://efaidnbmnnnibpcajpcglclefindmkaj/https://www.cig.gov.pt/wp-content/uploads/2017/08/guiao-plano-igualdade-empresas.pdf.

Comissão para a Igualdade no Trabalho e no Emprego, União Geral de Trabalhadores, Confederação Geral dos Trabalhadores Portugueses, Centro de Estudos para a Intervenção Social, Confederação do Comércio e Serviços Portugal, Associação Portuguesa de Ética Empresarial, Rádio e Televisão de Portugal, S. A., Instituto Superior de Ciências do Trabalho e da Empresa, & Instituto de Apoio às Pequenas e Médias Empresas e à Inovação. (2008). *Guia de Auto-Avaliação da Igualdade de Género nas Empresas*. Retrieved 4 November, 2023, from https://www.een-portugal.pt/info/RSO/Paginas/Apresentacao/indice.html.

Constituição da República Portuguesa (1976). Retrieved 14 November, 2023, from chrome-extension://efaidnbmnnnibpcajpcglclefindmkaj/https://www.parlamento.pt/Parlamento/Documents/CRP1976.pdf.

Constituição da República Portuguesa (2005). Retrieved 14 November, 2023, from chrome-extension://efaidnbmnnnibpcajpcglclefindmkaj/https://www.parlamento.pt/ArquivoDocumentacao/Documents/CRPVIIrevisao.pdf.

Dahlsrud, A. (2008). How corporate social responsibility is defined: An analysis of 37 definitions. *Corporate Social Responsibility and Environmental Management, 15*(1), 1–13. https://doi.org/10.1002/csr.132

de Almeida, P. (2022). Ética e Responsabilidade Social nos Negócios. *Comunicação & Inovação, 3*(5), 57–58.

Diretiva 2006/53/CE de 5 de julho do Parlamento Europeu e do Conselho, EUR-Lex: OJ L 204/23 (2006). Retrieved 13 April, 2022, from www.eur-lex.europa.eu/.

Dobers, P. (2009). Corporate social responsibility: Management and methods. *Corporate Social Responsibility and Environmental Management, 16*(4), 185–191. https://doi.org/10.1002/csr.201

Eriksson, P., & Kovalainen, A. (2008). Qualitative Methods in Business Research. *SAGE Publications Ltd*. https://doi.org/10.4135/9780857028044

Eurofound. (2023). *Igualdade de Género*. Retrieved 5 October, 2023, from https://www.eurofound.europa.eu/pt/topic/igualdade-de-genero?page=%2C4%2C34%2C0%2C0%2C0.

Eurostat. (2019). *Gender Employment Gap in the EU*. Retrieved 4 October, 2023, from https://Ec.Europa.Eu/Eurostat/Web/Products-Eurostat-News/-/EDN-20190307-1.

Eurostat. (2021). *Gender pay gap statistics*. Retrieved 4 October, 2023, from https://Ec.Europa.Eu/Eurostat/Statistics-Explained/Index.Php?Title=Gender_pay_gap_statistics.

Freeman, R. E. (2010). *Strategic management: A stakeholder approach*. Cambridge university press.

Froes, C., & de Melo Neto, F. P. (2004). *Gestão Responsabilidade Social Corporativa. O Caso Brasileiro*. Qualitymark.

Fundação Instituto de Administração. (2019, August 29). *Responsabilidade Social: o que é, importância e exemplos*. Retrieved 30 October, 2023, from https://Fia.Com.Br/Blog/Responsabilidade-Social/.

International Labour Organisation. (2009, June 23). *The ILO and corporate social responsibility*. Retrieved 30 October, 2023, from https://Www.Ilo.Org/Wcmsp5/Groups/Public/---Ed_emp/---Emp_ent/---Multi/Documents/Publication/Wcms_116336.Pdf.

Kroon, B., & Paauwe, J. (2014). Structuration of precarious employment in economically constrained firms: The case of Dutch agriculture. *Human Resource Management Journal, 24*(1), 19–37. https://doi.org/10.1111/1748-8583.12024

Lei Constitucional n.º 1/2005 de 12 de agosto da Assembleia da República. Diário da República: I série, No 155 (2005). Retrieved 10 April, 2022, from www.dre.pt.

Lei n.º 7/2009 de 12 de fevereiro da Assembleia da República. Diário da República: Série I, No 30 (2009). Retrieved 13 April, 2022, from www.dre.pt.

Michaels, E., Handfield-Jones, H., & Axelrod, B. (2001). *The War for Talent*. Harvard Business Press.

Neto, J. (2010). Ética, Responsabilidade Social e Gestão da Informação nas Organizações. *Informação & Sociedade: Estudos, 20*(3), 27–38.

Santos, E. (2022). Igualdade de género no trabalho: a realidade portuguesa. Retrieved 10 April, 2022, from https://Www.e-Konomista.Pt/Igualdade-de-Genero/.

Turker, D. (2018). Managing Social Responsibility. *Springer International Publishing*. https://doi.org/10.1007/978-3-319-91710-8

Universidade Europeia. (2023). *Ética e Responsabilidade Social*. Retrieved 10 October, 2023, from chrome-extension://efaidnbmnnnibpcajpcglclefindmkaj/https://www.europeia.pt/docs/codigo_etica_PT_27072023.pdf.

Index

A
Abolition, 233
Access to employment, 233
Adaptive work culture, 118
Affective commitment, 71, 73, 78–88
AI-amplified embodiment, 108
AI applications, 38, 39, 42, 61, 63–65, 67
AI-based solutions, 45
AI-driven technologies, 38
AI integration, 108
AI-powered technologies, 38
AI technologies, 37, 39, 46–48, 53, 57, 64, 66, 96, 120
Applications, 38, 39, 41, 42, 46–49, 52, 58–60, 63, 64, 67
Artificial General Marketing Intelligence (AGMI), 108, 121
Artificial Intelligence (AI), 37–43, 45–47, 49–67, 95, 97, 104, 106, 108, 120–122
Artificial intelligence technologies, 3
AR/VR integration, 107, 109
Ashby's Law, 204, 205, 210, 218–220, 224
Attracting talent, 151
Attractive career paths, 151
Augmented and Virtual Reality (AR/VR), 95, 96, 104, 106–109
Augmented and Virtual Reality (AR/VR) interface, 108
Augmented Reality (AR), 95, 97, 104, 106–108, 120, 122
Augmented Reality/Virtual Reality/ Artificial Intelligence (AR/VR/AI), 95, 97, 104, 106, 122
Automation, 41, 43, 57, 66, 111, 112
Autonomous living, 116
Autonomous systems, 38
Awards, 233, 239

B
Balanced, 171
Banking, 41, 42
Beer's VSM, 207, 213
Benefits, 231, 233
Biases in recruitment, 62
Bills, 39
Building, 22, 75, 100, 106, 114, 119, 127, 133, 140, 163
Business, 1, 4–6, 9, 17, 29, 31, 37–45, 47, 55, 60, 64, 66
Business activities, 208
Business activity, 158
Business ethics, 127–132, 138, 140, 141
Business impact, 157
Business landscape, 96, 116
Business management, 148, 152
Business strategies, 129

C
Cambridge analytica, 128, 129
Capitalization of resources, 229, 236
Career, 170, 172, 174–176, 178–185
Career management, 229, 230, 236, 239
Career progression, 178, 179
Career protection, 233
Challenges, 37, 49, 56, 59, 62, 64, 67, 148, 151, 153, 162, 164
Changes, 169, 170, 172–176, 180–182
Child labor, 233
Citizenship behaviors, 170, 172

C. Machado (ed.), *Building the Future with Human Resource Management*, Management and Industrial Engineering, https://doi.org/10.1007/978-3-031-52811-8

Climate supporting responsible management, 135
Cloud-based services, 57
Cluster, 202, 219
Cluster generation, 8
Cluster-specific conditions, 202
Codes of conduct, 161
Cognition, 191, 193–195, 198, 200, 202, 214, 219, 222, 224
Cognition of design, 191–224
Collaboration, 107, 110–112, 115, 116, 118–121
Collusion, 6
Commitment, 170, 172, 173, 178, 180, 183–185
Communication, 229, 230, 237, 240
Communication channels, 138
Companies, 147–150, 152–165, 229–242
Company green performance, 77, 78
Company innovation infrastructure, 202
Company innovation orientation, 202
Compensation and benefit, 46, 52, 53
Competitors, 148
Complexity, 95–99, 102, 104–106, 110, 111, 113, 120, 122
Compliance, 149, 151, 153, 159
Computational technologies, 39
Continuous feedback mechanism, 55
Continuous innovation, 4
Continuous training, 156
Coordination, 198, 208, 209, 214
Corporate citizenship, 127, 130
Corporate image, 152
Corporate shared value, 118
Corporate Social Responsibility (CSR), 128–130, 137, 148, 149, 152, 153, 157, 165, 230–233
Corporate volunteering, 154
Create value, 154
Creation of work, 231
Creative aspects, 100
Creative atmosphere, 115
Creativity, 99, 104, 111–113, 115–117, 122
Culture, 99, 104, 110–112, 115, 117, 118
Customer analysis, 40
Customers, 148, 150, 158, 162–164
Cybernetics, 95, 97, 103, 104, 118–120, 122, 191, 193–195, 197–199, 201, 202, 204, 207, 211–214, 218, 222, 224
Cyber-physical, 96, 105–107, 110, 111, 115, 120, 122
Cyberphysical Navigator Model, 95, 96, 121
Cyberphysical Networked Marketing (CNM), 120–122
Cyber-physical reality, 95–97, 104–108, 116–118, 120, 121
Cyber security, 40
Cybersecurity, 40, 63

D

Data-driven decisions, 121
Data-driven insights, 38, 61
Data scandal, 128, 129
Data workbench, 58
Daunting task, 151
Decision-making, 38, 45, 46, 48, 50–52, 56, 60, 62, 64, 65, 67
Decision-making processes, 238
Deep Learning (DL), 38, 108
Define, 114
Definition, 147, 149, 151
Design, 99–101, 109, 111, 120, 191–206, 210, 213–215, 218–220, 222–224
Design field, 193, 195
Design for complexity, 219
Designing complexity, 218
Design of cognition, 191–224
Design thinking, 197, 200
Design Thinking (DT), 95, 99, 101, 102, 104, 111–113, 120, 122
Design thinking principles, 96
Design thinking process, 112–115
Design WeltAnschauung (DWA), 98, 99, 102–104, 122
Development, 3, 5–10, 15, 30
Development plans, 233
Digital age, 104, 106, 112, 113
Digital HRM, 38
Digitalization, 95, 96, 104, 111, 112
Digital world, 37, 59
Digital world of work, 59
Dignity, 229, 233, 236, 240
Direction, 209–211, 219
Discipline Management Systems (DMS), 54
Discrimination, 233, 235, 238–240
Disruptive landscape, 117
Diversity, 154, 159, 161, 162
Duties, 179–184
Duty of respect, 236, 240

E
Eco-friendly, 72, 73, 75
Eco-friendly behaviours, 72, 73, 75
Ecological balance, 128, 129, 140
Efficient onboarding, 61
Elimination, 233
Embedded, 196, 202, 206, 207, 218, 219, 224
Embedded artificial intelligence, 58
Embedded environment, 104
Embodiment, 191, 193–195, 200–202, 205, 215, 218–220, 224
Emerging skills, 51
Empathy, 111, 115
Employee engagement, 46, 54–56, 58, 61, 63
Employee experiences, 18, 23
Employee involvement, 154
Employee job satisfaction, 156
Employee-management relations, 154
Employee performance, 149
Employee performance evaluation system, 75
Employee relations, 53
Employee retention, 55, 56
Employees, 230–233, 236–239, 241, 242
Employees experience, 66
Employees'green behaviour, 73, 76, 77, 82, 87
Employee / student wellbeing, 109
Employee training, 154, 155, 159
Employee well-being, 96, 109, 115, 118, 120
Employer-employee relationship, 184
Employment, 169–175, 178–181, 183–185, 233
Employment relationship, 233
Engagement, 149, 158
Enhancing, 108, 218
Environment, 95, 96, 98, 103, 105, 107, 108, 110–112, 116–122, 192, 194, 195, 199–207, 209, 211, 214, 216–220, 222–224, 231, 232
Environmental friendly business, 135
Environmental health in production, 135
Environmental management, 135
Environmental performance, 71–75, 77, 78, 83, 84, 86, 87
Environmental sustainability, 98
Equality, 230, 233–236, 238–240, 242
Equal working conditions, 234
Ethical behaviors, 130, 133–135, 138
Ethical codes, 130–132
Ethical committees, 130
Ethical conduct, 133
Ethical culture, 133, 134, 136–138, 141
Ethical issue, 64
Ethical leaders, 133, 134, 137, 140
Ethical leadership, 132–134, 136, 137, 140, 141
Ethical norms, 173
Ethical organizational culture, 135, 136
Ethical principles, 128, 130, 132–135, 138, 139
Ethics, 129, 130, 132–134, 139, 232, 237
European context, 234
Evaluation of workers, 156
Evolution, 38, 44, 45
Exclusion list, 12
Executive direction, 211
Expansion, 38, 63
External communication, 148, 158
External equity, 150, 151
External level, 236, 242

F
Fair work distribution, 151
Family, 229, 230, 234–237, 240, 241
Family assistance, 229, 230, 237, 241
Finance, 41, 42
Firm, 202, 206, 210, 211, 218–220, 223
First suit, 57, 59
Flexible work, 116–118
Focused HRM practices, 137
Followers, 132–134, 136
Formal, 230, 233, 240
Formation of local clusters, 119, 120
Framework, 95–106, 108, 110, 112, 113, 117, 118, 120–122
Fraud detection, 41, 42
Functional interplay, 106
Functional managers, 151
Future, 198, 201, 210, 211, 229, 241

G
Gender, 234, 235, 238–240, 242
Gender equality, 229, 230, 233–240, 242
GHRM practices, 77, 82–85
Global crises, 97
Go green, 71–91
Green awareness, 74
Green behaviour, 71, 73, 75–77, 79–89
Green employer branding, 74
Green Human Resources Management (GHRM), 71–74, 76–89

Green Incentive and Remunerations System (GIRS), 75
Green management of human resources, 72
Green performance, 72, 75, 78–89
Green, performance, 79
Green Performance Management and Evaluation (GPME), 75
Green Recruitment and Selection (GRS), 72, 74
Green supply chain transformation, 140
Green training and development, 72, 74, 78
GRI indicators, 159
GRI standards, 154, 159

H
Happiness, 240, 242
Healthcare, 38, 42
Healthy working environment, 151, 233
High-performance practices, 147, 154, 155
High-performance work systems practices, 154
High-Tech Self-Provisioning (HTSP), 116
Holistic model, 96
Holistic nature, 195
Holistic perspective, 210, 219
HR management, 158
HR managers, 50, 51, 58
HRM innovation, 1, 3, 29, 30
HRM practices, 50, 59, 60, 63–65, 137, 138
HR professionals, 37, 44, 45, 47–50, 52, 53, 57–61, 63–65, 67
Humanities, 192
Human Resource Information Systems (HRIS), 45, 59
Human resource management practices, 232, 233
Human resource management processes, 151
Human resources, 147–149, 151–154, 157–159, 161–164
Human resources department, 147, 150, 151
Human resources management, 71–73, 170, 183
Human Resources Management (HRM), 1–5, 9–26, 29–31, 37, 43–46, 50–52, 56–60, 62–64, 67, 185, 233
Humans, 99, 100, 102, 106, 109, 111, 115, 122

I
Icosahedron, 215, 216, 218–220, 224
Ideate, 114
Ideological obligations, 173
Ideological PC, 169, 172, 173, 182, 184, 185
IGen, 238, 242
Image, 229, 230, 235–237, 240, 242
Increased risks, 63
Incremental innovation, 4
Individual interests, 174, 184
Individual staff green behaviours, 75
Information, 229, 230, 237, 239, 240
Information and Communication Technologies (ICT), 59
Information integration, 140
Information mapping, 8
Information technology, 3
Innovation, 1–7, 10–13, 16, 18, 19, 23, 26, 29–31, 96, 98, 99, 101, 107, 110, 112, 115, 118, 122, 202, 205, 211
Innovation analysis, 7, 11, 12
Innovation capability, 5
Innovative, 98, 103, 110, 111, 113–115, 117, 121, 122
Intangible assets, 148
Integrated Viable System Model (iVSM) model, 106
Integrate sustainability, 129, 136
Integration, 232
Intelligent organization, 108
Inter, 54
Interact, 206, 212
Interdisciplinary team, 114–116
Internal, 148, 151, 155, 158, 160, 161
Internal level, 236
Internal operations, 209
International Organization for Standardization (ISO), 232
Interorganizational relationship management, 140
Intersectional perspective, 235
Intrapersonal skills, 54
Invention, 2, 6
Inverse Document Frequency (IDF), 13
Invoices, 39
IPC system, 22, 24
IT development, 9

J
Job security, 155, 156, 169, 170, 172, 173, 175, 178–181, 184

K
Knowledge, 154, 158

Knowledge management, 5

L
Labor policies, 233
Labor practices, 233
Labor rights, 233
Lacks human touch, 63
Landscape analysis, 1, 8
Law 11/2018, 159
Learning, 5, 8, 179
Learning and development, 55, 61
Lemmatization, 12
Lifelong learning, 118, 229, 230, 236, 239
Logic of design, 192, 194, 197, 199
Loyalty, 170, 171, 173, 179, 181, 185, 229, 236

M
Machine-generated errors, 64
Machine learning, 38, 39, 41, 46–48, 61
Machine learning algorithms, 38, 39, 41, 47, 48, 61
Machine Learning (ML), 106, 108
Machines, 106
Mainstreaming, 235
Management, 193, 200–202, 204–207, 209, 211, 217–219, 223
Management cybernetics, 96, 109
Management practices, 129, 132, 141
Management unit, 110, 111, 120, 121
Managerial approach of applying design, 101
Man-machine-interface, 95, 96, 104, 106, 108, 122
Marketing, 95–98, 102, 108, 118–121
Marketing strategy, 108, 121
Maternity, 229, 230, 237, 240, 241
Mature, 238
Men, 230, 234–236, 238–241
Microservices, 57, 58
Military, 169, 170, 173–176, 178–185
Military context, 169, 173, 181, 183, 185
Military organization, 169, 170, 174, 176
Mission, 229, 230, 236–238
Modifying products, 120
More responsible, 136, 137, 140, 141, 157, 229
Motivation, 229, 236
Multiple discrimination, 235

N
Natural Language Processing (NLP), 38, 48
Necessary synergy, 201
Neural networks, 38, 46
New challenges, 162
New work, 95, 104, 111, 112, 115–120, 122
New work approaches, 96
NLP capabilities, 48
Non-discrimination, 153, 154, 159, 233–235, 240, 242
Non-financial reporting, 159
Now, 209–211, 214, 217
Number of states, 205

O
Observe, 103, 114
Occupation, 233
Onboarding, 46, 49, 51, 57, 59–61
Operational integration, 140
Operations, 204, 206, 207, 209, 211
Opportunities, 37, 38, 47, 51, 53–55, 58, 59, 62, 150, 154, 155, 158, 159, 161, 162, 164
Organisational behaviour framework, 132
Organisational structure, 109
Organisational unit, 105, 106
Organizational charts, 212
Organizational cognition, 218, 220
Organizational culture, 132, 133, 135, 136, 140, 238, 240
Organizational goals, 175, 180, 183, 185
Organizational performance, 107, 121
Organizational resilience, 96, 97, 99, 103, 104, 106, 109, 120, 121
Organizational restructuring, 173, 174, 185
Organizational strategies, 138
Organizational structure, 95, 96, 98, 99, 105–107, 110, 112, 117
Organizational success, 23
Organizations, 37, 39, 43–56, 59–67, 71–79, 83, 86–89, 229–233, 236, 239, 240, 242
Organization‘s identity, 103
Organization system, 206
Outside, 205, 207, 210, 213
Own decision-making, 156

P
Parental leave, 235, 241
Participating, 120
Participation, 118
Participation in the results, 156

Participation of workers, 229, 230, 236, 240
Patent analysis, 3–10, 31
Patent families, 1, 3, 7, 11–24, 29, 30
Patent landscape, 1, 3, 7, 8, 10–12, 14, 15, 26, 30, 31
Patents, 1–3, 5–16, 18, 19, 21, 23, 25, 26, 29–31
Patent system, 6
Paternity, 229, 230, 237, 240, 241
People, 1, 3, 6, 10, 11
People management, 184
Perceptiveness, 54
Performance, 169, 171, 173, 175
Performance evaluation, 233
Performance management, 45, 46, 51, 52, 60
Personalization, 41, 53
Personal life, 229, 230, 235–237, 240, 241
Personnel, 229, 230, 236, 238
Phenomenological approach, 199
Phenomenology, 193, 197, 198, 203
Philosophical foundations, 197
Policies, 128–130, 132, 137–139, 141
Portugues context, 235
Portuguese army, 170, 175, 176, 180, 181, 184
Post-covid, 96, 103, 107
Practical problem-solving approaches, 101
Practices, 37, 44, 46, 47, 51, 53, 56, 58–61, 63, 64, 67, 127–130, 132, 134, 137, 138, 140, 141, 232, 233, 236, 242
Pragmatism, 193, 197, 199, 200
Predictive maintenance, 41
Prestige, 229–231
Principle of recursion, 201, 202
Processes, 148–150, 156, 158, 162
Productivity, 229, 236, 237
Pro-environmental behaviours, 75
Professional activity, 234
Prohibition of discrimination, 233
Prohibition of harassment, 234
Promoting, 232, 235–237
Promoting transparency, 151
Promotion of gender equality, 229, 230
Protection, 229, 230, 232, 235–237, 240, 241
Prototype, 101, 115
Psychological Contract (PC), 169–173, 175, 176, 178, 180–185
Psychological Contract (Performance), 169

R

Radical constructivism, 200, 201, 213
Rapprochement, 102
Recognition, 233, 236, 241, 242
Reconciling professional, 234, 235, 237, 241
Recruitment, 229, 230, 233, 236, 238
Recruitment and selection, 45–48, 60, 64
Recruitment based on responsibility, 156
Recruitment processes, 234, 238, 239
Redefining productivity, 120
Reduced wage labour, 116
Reduction, 235
Reimagining work, 115
Reinvention of HRM, 45
Relational, 169–173, 175, 178, 180, 181, 183, 184
Relational dimension, 181
Relationships, 231–233, 236
Remuneration, 150, 151, 154–156, 159, 229, 230, 233, 234, 236, 239
Repetitive tasks, 41
Representative organizations, 236, 240
Reputation, 229–231, 235, 236
Requisite variety, 200
Research organizations, 6
Resilience, 95, 96, 98, 99, 102, 107, 112
Resilience of organizations, 97
Responsibility, 127–132, 134, 136–140
Responsibility projects, 127
Responsible companies, 147, 151, 153, 161, 164
Responsible employer rankings, 160
Responsible human resource management, 147, 149, 151, 153, 155, 158, 165
Responsible management, 127–132, 133, 135, 137, 139–141, 148, 154, 155, 157, 158, 162, 165
Restructuring, 169, 170, 176, 185
Retail, 43
Retaining talent, 151
Retention, 46–48, 55, 57–59, 61
Revolutionary advancement, 38
Revolutionizing industries, 38
Rights, 173, 178–181, 183, 184
Robotic Process Automation (RPA), 57
Robotic technology, 3

S

Salary remuneration, 151
Science, 192, 193, 195, 196, 198, 200–204, 213, 218, 222, 224
Scientific information, 6
Second-order cybernetics, 191, 193, 194, 213

Selection, 229, 230, 233, 236, 238, 239
Self-determined work, 117
Self-monitoring, 156
Self-motivation, 156
Self-organization, 203, 204, 214, 218
Self-reflection, 54
Sex, 233–235, 240
Shared value, 98, 104, 117–119
Short-term profit, 98, 103
Single, 233
Six Forces Model (SFM), 218–221, 224
Skill set, 50, 64
Social dialogue, 229, 230, 236, 238, 240
Socially responsible companies, 148–150, 162
Socially Responsible Human Resources Management (SRHRM), 137, 138, 232, 233
Socially responsible practices, 147, 151, 153
Social media, 37, 39–41, 56, 60
Social performance indicators, 159
Social responsibility, 98, 117–119, 148, 149, 151, 152, 165, 229–233, 236, 238, 242
Society, 95, 96, 103, 104, 107, 118–120, 148, 158, 159, 162, 164
Software tools, 57
Solid foundation, 109
Spillovers, 6
Staff, 71–80, 83, 85–87
Stakeholders, 130–132, 134–137, 139–141, 148, 149, 158, 161, 230, 231, 233
Stemming, 12
Strategic approach, 148
Strategic choices, 150
Strategic decisions, 150
Strategic development, 8, 233
Strategic direction, 210
Strategic foresight, 211
Strategic integration, 140
Strategic management, 180
Strategy models, 221
Streamline processes, 5
Structural components, 129
Structuring business ethics, 132
Suppliers, 148, 164, 165
Supply Chain Ethical Leadership (SCEL), 140
Supporting, 208
Sustainability, 72, 75–77, 87, 127, 128, 130–132, 135–141
Sustainable development, 127, 128, 130, 131
Sustainable framework, 95–122
Sustainable process, 75
Sustainable supply chain management, 139
System, 193, 194, 201–212, 214, 218, 223
System 1, 207–209
System 2, 207–209
System 3, 207, 209–211
System 4, 210, 211
System 5, 211

T

Talent management, 233
Team Syntegrity Model (TSM), 214–217, 219, 220, 224
Teamwork, 9, 10
Technical, 238, 239
Technical field, 3, 9, 10, 15, 29
Technological advancement, 1, 2, 5, 8, 16, 21, 29
Technological development, 235
Technological strategies, 10
Technological trends, 7–11, 31
Technology, 37, 38, 40, 42, 44–48, 50, 52, 54, 58, 64, 66, 67
Technology estimation, 10
Term Frequency-Inverse Document Frequency (TF-IDF), 13, 30
Term Frequency (TF), 13
Test, 115, 119
Timeframe, 103
Tool, 202, 218
Topic identification, 8
Train employees, 151
Training, 138, 231, 233, 234, 237, 239
Training and development, 46, 50, 51, 61
Transactional, 171–173, 175, 179, 183, 184
Transactional exchanges, 173
Transformative leaders, 133
Transitional, 169, 171, 172, 181–183
Transitional tendency, 181
Transportation, 43
Transversal skills, 238
Triple bottom line, 129

U

Understand, 103, 107, 111, 114, 117
User-friendly work environment, 37

V

Values, 229, 230, 232, 236–239
Variety, 194, 195, 200, 202, 204, 205, 218, 220
Viable System Model (VSM), 106, 109, 110, 203, 207–212, 214, 215, 220, 224
Virtual reality, 46, 50
Virtual Reality (VR), 97, 101, 104, 106–109, 120, 122
Vision, 231, 232, 237, 238
Vitality, 10
VSM's system, 209

W

Wage discrepancy, 235
Waves, 44, 45
Web search, 162
Weekly working hours, 155
Well-being, 232, 233, 240, 242
Wider environment, 219
Women, 230, 234–236, 238–240
Women's illiteracy rates, 235
Work, 229, 232, 233, 235, 238, 239, 241
Work environment, 37, 112
Workers, 148, 151, 153, 154, 156, 162–164
Working hours, 155
Working hours of employees, 151
Workload commensurate, 151
Workplace, 169, 236, 240
Workplace behaviors, 135
Workplace harassment, 239, 240
World Health Organisation (WHO), 162

MIX
Papier aus verantwortungsvollen Quellen
Paper from responsible sources
FSC® C105338

If you have any concerns about our products,
you can contact us on
ProductSafety@springernature.com

In case Publisher is established outside the EU,
the EU authorized representative is:
Springer Nature Customer Service Center GmbH
Europaplatz 3, 69115 Heidelberg, Germany

Printed by Libri Plureos GmbH
in Hamburg, Germany